TABLE OF CONTENTS

Page

LIST OF ILLUSTRATIONS ii

LIST OF TABLES iii

LIST OF ABBREVIATIONS iv

CHAPTER

    1. INTRODUCTION 1

    2. LITERATURE REVIEW 14

    3. RESEARCH METHODOLOGY 48

    4. ANALYSIS 58

    5. CONCLUSIONS AND RECOMMENDATIONS 77

APPENDIX

    A. COMPARISON OF ENGINEER COMPANY MTPS 86

    B. INTERVIEW QUESTIONS AND RESULTS 87

    C. CRITICAL TASK LISTS FOR 215D AND 81T SOLDIERS 99

    D. ESSENTIAL TASK LIST FOR 21B OFFICERS 105

GLOSSARY 111

REFERENCE LIST 114

LIST OF ILLUSTRATIONS

Figure	Page
1. METL Development Process	6
2. SKB Selection Process	7
3. Geospatial Engineering Roles	9
4. Essential SKB Selection Progress	57
5. Essential SKB Distribution by Officer Rank	59

LIST OF TABLES

Table	Page
1. Terrain Analysis Tasks	51
2. Data Collection Tasks	52
3. Data Generation Tasks	53
4. Database Management Tasks	53
5. Data Manipulation and Exploitation Tasks	54
6. Cartographic Production and Reproduction Tasks for Engineer Officers	54
7. Geodetic Survey Tasks for Engineer Officers	54
8. Terrain Advice Tasks for Engineer Officers	55
9. Navigation Tasks	56
10. Comparison of Processes and Functions	64
11. Essential Tasks for 21B Officers, 215D Warrant Officers, and 81T Soldiers	67
12. Training Strategy for Untrained SKBs	82

LIST OF ABBREVIATIONS

ABCS	Army Battle Command System
BOLC	Basic Officer Leaders Course
CGSC	Command and General Staff College
CTC	Combat Training Center
DTSS	Digital Topographic Support System
ECCC	Engineer Captains Career Course
FBCB2	Force XXI Battle Command Brigade and Below
FM	Field Manual
GEODDUC	Geospatial Digital Data User's Course
GI	Geospatial Information
GI&S	Geospatial Information and Services
MCS	Maneuver Control System
MDMP	Military Decision Making Process
METL	Mission Essential Task List
MOS	Military Occupational Specialty
MTP	Mission Training Plan
OC	Observer Controller
OES	Officer Education System
ROTC	Reserve Officer Training Corps
SKB	Skills, Knowledge, Behaviors
TOMC	Topographic Officers Management Course
USMA	United States Military Academy

CHAPTER 1

INTRODUCTION

> We are not fit to lead an army on the march unless we are familiar with the face of the country--its mountains and forests, its pitfalls and precipices, its marshes and swamps. We shall be unable to turn natural advantages to account unless we make use of local guides. (2002, 65)
>
> Sun Tzu, *The Art of War*

As noted over two thousand years ago, military leaders must know the terrain they will operate. Sun Tzu advocated the use of local guides to take advantage of the face of the country. In the Army, engineer officers are the local guides who assist commanders to see the terrain of the battlefield. But what does it take to be the terrain guide, or expert, for commanders at tactical levels of the Army? This research study investigates past examples and current and future force requirements for the essential set of skills, knowledge, and behaviors (SKBs) that engineer officers must possess to be terrain experts.

Terrain deals with all the physical and cultural geographical features of a given area (Collins 1998, 404). The study of terrain, or topography, has been an enduring combat responsibility of engineers throughout US Army history. From 1838 to 1863, engineer officers were divided into the Topographic Engineers ("topogs") and the Corps of Engineers. Several noted military leaders were surveyors or engineers: George Washington, Robert E. Lee, Henry W. Halleck, and George McClellan, while others were topographers as lieutenants and captains: George G. Meade, Joseph E. Johnston, John Pope, and John C. Fremont (Traas 1993, 6). Other famous generals, such as Stonewall Jackson, George S. Patton, and Douglas MacArthur, had their own personal team of topographic engineer officers to create maps for their planning. Engineer Captain Jedediah Hotchkiss served as the terrain advisor to generals Jackson, Early, Ewell, Lee, and Garnett of the Confederate Army in the Civil War. In the introduction to his book, *Make Me a Map of the Valley*, Hotchkiss is described:

> The engineer, with his quick perception of terrain could swiftly supply accurate sketches to the general [Jackson], who had no real facility for grasping the lay of the land. . . . Before movements of the army he was frequently called in to give advice on the terrain. He made sure that he was able to furnish graphical representations of any point on which Jackson was not clear. (1973, xxi)

Terrain experts like Captain Hotchkiss aid their respective commanders in exploiting the terrain to achieve victory.

During the Civil War, the Topographic Engineer branch was reunited with the Corps of Engineers, where it remained an integral, though specialized, skill. In World War II, topographic engineer battalions and companies supported every army and corps organization. Since then, engineer officers specialized in one of three specialty codes (SCs) or Military Occupational Specialties (MOS): Engineer (MOS 21), Topographic Engineer (MOS 22), and Construction Engineer (MOS 23) (Reminger 1983, 35). The Topographic Engineer MOS 22 was later changed to MOS 21C in the early 1990s, but essentially remained a separate career path from the combat and construction engineer MOSs. In 1984, the combat and construction engineer officers MOSs combined into the MOS 21B Combat Engineer series. In 1996, the combat, construction and topographic engineer officers MOSs combined--all engineer officers became MOS 21B Combat Engineers capable of topographic and construction engineering. As of 2002, geospatial engineering is the new term replacing topographic engineering in Army doctrine (Aadland and Allen 2002, 8). Geospatial engineering is:

> The collection, development, dissemination and analysis of positionally accurate terrain information that is tied to some earth reference, to provide mission tailored data, tactical decision aids and visualization products that define the character of the zone for the maneuver commander. (FM 3-34 2003, 4-9)

In a sense, an engineer officer is a "jack of all trades" in engineering, but he may not be a master of all, especially when it comes to terrain expertise. In the article "We're All Terrain Experts," Major David Treleaven voiced his concern as a combat engineer taking on this additional role.

> Engineer officers are expected to be terrain experts. . . . My personal frustrations and shortcomings point to a training deficiency that must be addressed before we can adequately label ourselves as both terrain experts and topographic officers. (1995, 8)

The evidence from combat training centers (CTCs) shows mixed results on how well engineer officers are mastering geospatial-related tasks that are important in planning and preparing operations during the military decision making process (MDMP). CTC observations are published in the Center for Army Lessons Learned (CALL) bulletins, such as to following observation made at the National Training Center (NTC) in 1999:

> The brigade staff does not appreciate the significant impact that terrain may have on their units' operations. Many ABEs [assistant brigade engineers] brief terrain only in general terms (mountain high, valley low) and do not discuss OCOKA [obstacles, cover and concealment, observation and fields of fire, key terrain, avenues of approach] or effects on trafficability. (McGinley 1999, 78)

Yet, there are success stories of engineers meeting the geospatial engineering challenge. For example, in 1997 the then Major General Leon J. Laporte, Commanding General of the 1st Cavalry Division, highlighted the significant impact that good terrain expertise made to his division during several real deployments to Korea and training exercises at NTC. "The production and interpretation of . . . terrain products gave the 1st

Cavalry Division and its leaders the confidence to plan and execute its mission in some of the toughest terrain we might be asked to fight upon in the future" (Laporte and Melcher 1997, 76).

Determining what geospatial SKBs made a difference between the substandard terrain expertise at the CTC and the robust terrain support to the 1st Cavalry Division is important in order to improve the OES training for today's and tomorrow's engineer officers. No longer is topographic engineering delegated to a few modern day 'topog" officers. It is a cornerstone in the engineer regiment's relevance to the Army. It provides the common operational picture (COP) of the battlefield to attain assured mobility and battlespace information for countermobility, survivability and construction operations. It demands the attention of professionals armed with geospatial engineering skills, knowledge, and behaviors of the entire engineer team.

> Terrain is a source of friction in war, the engineers . . . are either the lubricant or the sand in that friction. To be an engineer requires a basic, fundamental understanding of terrain because we are the ones who will shape the battlefield. (Arnold 1997, 13)

Problem Statement and Research Questions

In order to meet the geospatial engineering challenge to the branch, this study seeks to answer the primary question: what skills, knowledge, and behaviors do engineers officers need to be terrain experts at the tactical levels of the Army? The Army includes three forces: the Current (or Legacy) Force, the Interim Force, and the Objective Force. The Current Force represents the majority of tactical units today. The Interim Force represents the modernization of the Current Force with available technologies, such as the Stryker Brigade Combat Team (SBCT). The Objective Force represents the concept force of the Army ten to twenty years from now. The primary question is further explored through four secondary questions.

1. What is an engineer officer terrain expert?

2. What terrain expertise was valuable or lacking in military operations prior to the engineer officer consolidation in 1996?

3. What are the terrain expertise requirements for engineer missions in the Current and Interim Forces?

4. What are the terrain expertise requirements for the Objective Force?

The first subordinate question defines what a terrain expert is based on current doctrine, professional articles, and future concepts. The remaining secondary questions serve as areas of investigation for the SKBs, or tasks, that are required, exemplified, or recommended from past, current, or future military activities and personnel.

Methodology

The research and selection process of essential SKBs flows similar to the mission essential task list (METL) development process as described in FM 7-0, *Training the Force*. The METL development process was chosen because:

1. It primarily focuses on selecting essential tasks at tactical levels taken from comprehensive review of wartime requirements. All company-level and above units must have METLs to focus their efforts.

2. It assists in establishing the hierarchy of supported and supporting tasks. All units have METL tasks supported by dozens of collective, leader and individual subordinate tasks.

3. It has served the Army well since 1991 in FM 25-100 and continues as the Army's METL process in FM 7-0.

In the METL development process (Figure 1), the commander and his staff research a wide variety of unit and mission related documents to establish what the critical war-time tasks are that the entire unit must accomplish. The availability of resources does not affect this process. Once a proposed METL is generated, the commander and his staff review and forward it for the next higher commander's approval. The commander then assesses his unit's readiness and builds his training plans based on the METL training objectives.

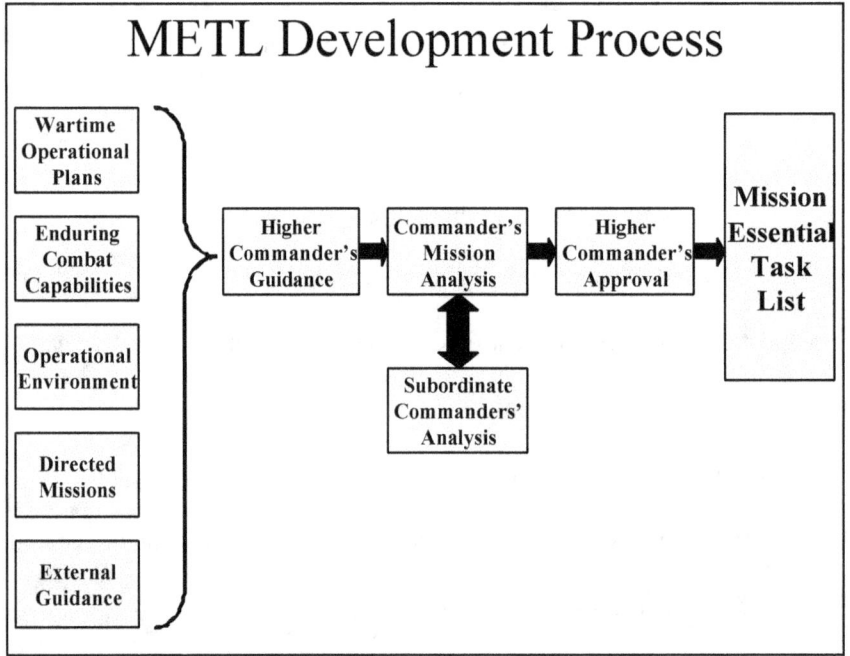

Figure 1. METL Development Process (FM 7-0 2002, 3-3)

In like manner, this research study develops an essential list of SKBs, or tasks, for the engineer officer terrain expert at tactical levels (Figure 2). Just as with the METL, the

SKBs are unconstrained statements of tasks required for the officer's mission. Chapter 2, Literature Review, examines a variety of Army, engineer, and geospatial related literature and field sources to extract potential mission essential SKBs. The tasks are grouped in accordance with the seven geospatial engineering functions as defined in the Army's Universal Task List (AUTL) in FM 7-15, and the two tactical applications of these functions: terrain advice and navigation. Chapter 3, Methodology, provides an assessment of the potential SKBs against five criteria to determine which tasks are essential. Chapter 4, Analysis, examines the essential SKB list to determine trends and gaps in the current training provided in the Officer Education System (OES). Chapter 5, Conclusions and Recommendations, offers several suggestions to assist engineer officers and the US Army Engineer School (USAES) develop, train, and execute the list of SKBs.

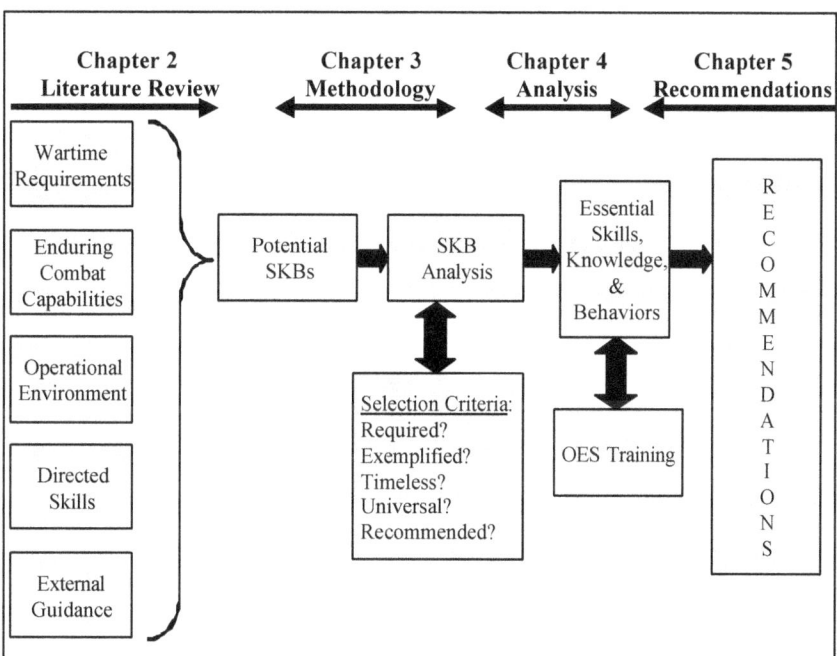

Figure 2. SKB Selection Process

Assumptions

In order to conduct this study, three assumptions are necessary. First, engineer officers are and will continue to receive sufficient training in combat engineering, general engineering, and the basics of tactical operations across the full-spectrum of offense, defense, stability, and support operations. As the USAES commandant, Major General Gill pointed out that engineer leader development must focus on supporting combined-arms warfighting, not just mastery of terrain visualization tools. Officers must have the composite engineering and tactical knowledge to effectively use terrain as an advantage (1996, 13). Second, the engineer officer must be able to accomplish the geospatial mission with "paper or plastic." The Current Force, especially in the reserve component, still depends on paper geospatial information and services (GI&S) products, manual analysis processes, and simple navigation equipment like the compass. The SBCT depends on plastic screen automated terrain analysis and visualization, such as the enhanced Digital Topographic Support System (DTSS) and Force XXI Battle Command Brigade and Below (FBCB2) system. Engineer officers must be able to exploit all forms of terrain data until future command, control, computers, communications, intelligence, surveillance, and reconnaissance (C4ISR) technologies are universally deployed. Third, engineer officers will remain generalists in all engineering missions versus returning to being specialists in the combat, general, or geospatial engineering domains through the development of the Objective Force.

Scope

This study uses the Army tactical task (ART) 1.1.1.5 "Conduct Geospatial Engineering Operations and Functions" in the AUTL to define the boundaries of what is meant by terrain in the label terrain expert. The functions include terrain analysis, data collection, data generation, database management, data manipulation and exploitation, cartographic production and reproductions, and geodetic survey (FM 7-15 2002, 1-6). Two significant applications of geospatial engineering, terrain advice and navigation, are also considered within the defined role of a terrain expert. Potential SKBs must fall within these nine functions and applications to be considered terrain expert tasks.

The study of engineer officers includes all ranks from second lieutenant to colonel and the primary tactical positions that engineer officers serve in the active and reserve components. Tactical positions include traditional leadership roles, such as platoon leaders, company commanders, and battalion training officers, and engineer-specific roles, such as task force engineers (TFEs), assistant brigade engineers (ABEs), and assistance division engineers (ADEs). The officer's role has overlap with that of the terrain-analysis technician (MOS 215D) and the topographic analyst (MOS 81T), but they also significantly differ. Figure 3 provides a comparison of the three roles.

Engineer Officer (MOS 21B)	The terrain visualization expert who assists the commander in visualizing the terrain, identifies and understands terrain aspects for exploitation by friendly and enemy forces, and provides subjective evaluation of the terrain's physical attributes and physical capabilities of vehicles, equipment and people.
Terrain Technician Warrant Officer (MOS 215D)	The terrain analysis and GI&S expert who helps the commander and staff understand the battlespace, assimilates and integrates large volumes of data, and transforms data into visualization information and knowledge.
Topographic Analyst Soldier (MOS 81T)	Performs cartographic and terrain analysis duties, collects and processes military geographic information from sensed imagery, digital data, and intelligence data, and advises the commander and staff on topographic operations and specialized product planning.

Figure 3. Geospatial Engineering Roles. (FM 3-34.230 2000, 3-1)

Since 215D warrant officers and 81T soldiers are found at the brigade, division, and corps levels, the study will consider them as available sources of expertise to engineer officers but not as the terrain experts directly responsible to the tactical commanders. Topographic engineer battalions are theater or operational assets and provide general support to tactical levels of command. This study will not critique the suitability, missions, or organization of the existing or proposed engineer units for the units of action (UAs) and units of employment (UEs) in the Objective Force.

Significance

This study concludes that there are specific, essential SKBs that engineer officers must master to be effective terrain experts in the Current and Objective forces. This research is significant because it is the first formal study of what an engineer officer is expected to know and to do as a terrain expert. The recommended SKBs provide a clear set of relevant tasks for leader development at USAES, in all engineer field units, and in professional self-development.

Definition of Key Terms

There are several important terms to define in regards to SKBs and terrain expertise. Additional key terms are located in the glossary. Carefully defined SKBs are essential to the Army's leader development program described in Department of the Army Pamphlet (DA PAM) 350-58 and now included in FM 7-0. They enable leaders to "successfully fulfill their roles, perform duties, and accomplish missions" (DA PAM 350-58 1994, 19). SKBs are gained during institutional training, practiced during operational assignments and refined during further self-developmental study and assessment. In this thesis, a skill, knowledge, or behavior is also called a task.

Skill. The ability to perform a job-related activity that contributes to the effective performance of a task (DA-PAM 350-58 1994, 92).

Knowledge. The minimum information about conditions, end results, means, and methods for reaching goals needed to ensure success in performing a task (DA PAM 350-58 1994, 91).

Behavior. An action or reaction to specific situations based on attitude, beliefs, and values. It is measurable and influenced by positive and negative reinforcement (DA PAM 350-58 1994, 90).

Task. A clearly defined and measurable activity accomplished by individuals and organizations. Tasks are specific activities that contribute to the accomplishment of encompassing missions or other requirements (FM 7-0 2002, G-17).

In the field of geospatial engineering, there are several commonly used terms dealing with military terrain. For the sake of simplicity, the term terrain expert is used instead of terrain visualization expert and geospatial expert. Terrain expert is defined in the next chapter as no doctrinal definition currently exists. Definitions for the seven-geospatial engineering functions from the AUTL are in the glossary.

OCOKA. A commonly used acronym and mnemonic for the military aspects of terrain. The acronym does not dictate the order in which the factors are evaluated. The aspects are observation and fields of fire, cover and concealment, obstacles, key terrain, and avenues of approach (FM 34-130 1994, G-8).

Terrain Analysis. The collection, analysis, evaluation, and interpretation of geographic information on the natural and man-made features of the terrain, combined with other relevant factors, to predict the effect of the terrain on military operations (FM 101-5-1 1997, 1-153).

Terrain Visualization. The process through which a commander sees the terrain and understands its impact on the operation in which he is involved. It is the subjective evaluation of the terrain's physical attributes as well as the physical capabilities of vehicles, equipment, and personnel that must cross over and occupy the terrain. Terrain visualization is a component of battlefield visualization. Engineers are responsible for providing the means to achieve terrain visualization (FM 3-34.230 2000, 1-5).

With respect to military operations in the Current force, Objective Force, the following terms describe various Army planning tools and automated computer systems used in today and tomorrow's tactical level Army organizations. Additional terms, such as Army Battle Command System (ABCS), battlefield operating systems (BOS), FBCB2, Joint Mapping Took Kit (JMTK), and Objective Force can be found in the glossary.

Assured Mobility. Actions that guarantee the force commander the ability to maneuver where and when he desires without interruption or delay to achieve his intent (FM 3-34.221 2002, 1-2).

Battlefield Terrain and Reasoning Awareness (BTRA). This allows friendly forces the ability to template potential obstacles and locations of where the enemy might place obstacles (Fowler and Johnston 2002, 13).

Digital Topographic Support System (DTSS). The topographic-engineer and topographic-analysis component of ABCS that provides critical, timely, accurate, and analyzed digital and hard-copy mapping products to the battle commander for terrain visualization. It is used by engineer topographic teams and companies (FM 3-34.230 2000, 4-5).

Engineer Battlefield Assessment (EBA). A parallel and companion staff action of the engineer to the S2's preparation of the intelligence preparation of the battlefield. It consists of three parts: terrain analysis, enemy mobility and survivability capabilities, and friendly mobility and survivability capabilities (FM 5-100 1996, A-2).

Intelligence Preparation of the Battlefield (IPB). A systematic approach to analyzing the enemy, weather, and terrain in a specific geographic area. It integrates enemy doctrine with the weather and terrain as they relate to the mission and specific battlefield environment (FM 101-5-1 1997, 1-84).

Military Decision Making Process (MDMP). A single, established, and proven analytical process consisting of seven distinct steps. It is an adaptation of the Army's analytical approach to problem solving that assists the commander and staff in developing estimates and a plan (FM 101-5 1997, 5-1).

Stryker Brigade Combat Team (SBCT). A brigade designed to optimize effectiveness and balance the traditional domains of lethality, mobility, and survivability with responsiveness, deployability, sustainability, and a reduced in-theater footprint. It is a full-spectrum combat force with the core qualities of mobility and decisive action through dismounted infantry assault (FM 3-34.221 2002, 1-1).

CHAPTER 2

LITERATURE REVIEW

Introduction

This chapter explores existing and potential SKBs found in a wide variety of primary and secondary reference materials and field sources. Many of the terrain-related tasks are common across the spectrum of past, current, and future events and doctrine. To describe these tasks as concisely as possible, the chapter is divided into three sections: a description of the sources cited, a proposed definition for terrain expert, and a listing of the potential SKBs. The final list of essential SKBs is in Appendix D.

Past, Current and Future Research Sources

Similar to the five sources of input in the METL development process, there are five sources of input for potential engineer officer SKBs. They are wartime requirements, enduring combat capabilities, operational environment, directed skills, and external guidance. These five areas, as with the proposed SKBs they provide, cross the spectrum of past, present and future operations and doctrine.

Wartime Requirements

Wartime requirements are the missions and tasks that units and individuals must fulfill to meet the Army's wartime operational and contingency plans. The Army's overall mission is stated in FM 1, *The Army*, and its METL is defined in FM 3-0, *Operations*. FM 7-15, *The Army Universal Task List* (AUTL) breaks down the Army's METL into ARTs organized under seven BOS elements. Most engineer unit METLs are derived from these ARTs. Engineers primarily focus on the BOS: Mobility, Countermobility, and

Survivability, yet the geospatial ART "Conduct Geospatial Engineering Operations and Functions" is found under the BOS: Intelligence (FM 7-15 2002, 1-6). Geospatial Engineering supports the higher ART of "Perform IPB" conducted at all tactical levels of command.

Engineers directly support the AUTL and FM 3-0 in their capstone document, FM 5-100 (soon to be FM 3-34), *Engineer Operations*. This manual makes it clear that terrain is an integral component of everything the branch brings to the Army:

> Engineers adapt terrain to multiply the battle effects of fire and maneuver. This engineer component of the close combat triad (fire, maneuver, and terrain) is described within the five engineer battlefield functions: mobility, countermobility, survivability, general engineering, and topographic engineering. (FM 5-100 1996, X)

To further apply doctrine, the Army uses Mission Training Plans (MTPs) to break down the BOS missions into supporting collective, leader, and individual tasks. Every Army unit, platoon-level and above, has a specific MTP. There are over one hundred different MTPs that cover the engineer branch. For this study, a review was conducted of the combat engineer company MTPs (mechanized, light, airborne, air assault, and corps wheeled) to find terrain-related leader tasks. The combat engineer company directly supports a maneuver commander, and combat engineers comprise approximately fifty-six percent of the engineer branch. The MTP for the combat engineer company in a heavy division, ARTEP 5-335-65-MTP, lists six primary missions supported by seventy supporting collective tasks. Sixteen of these seventy collective tasks require some level of terrain expertise from the company's engineer officers. These same sixteen collective tasks are also prevalent in the MTPs for the other combat engineer companies in light infantry, airborne, and air assault divisions as well (Appendix A).

The CTCs evaluate units against their respective ARTEP standards. CALL bulletins catalog the lessons learned from these evaluations, to include geospatial engineering lessons from the various unit training rotations. Ten semi-annual CALL bulletins since 1996 provide positive and negative feedback on performance of terrain analysis and terrain visualization by engineer officers and units at brigade level and below.

Enduring Combat Capabilities

Enduring combat capabilities are those unique contributions that engineers make to the Army so it can successfully accomplish its mission. The Corps of Engineers (engineer branch) has had three enduring capabilities since its formation in the early 1800s: combat engineering (mobility, countermobility, survivability), general engineering (construction and prime power), and topographic engineering (geospatial). Army Regulation (AR) 115-11, *Geospatial Information and Services*, specifies that engineers are the providers of geospatial support to the Army.

Several FMs further break down geospatial engineering from FM 5-100, *Engineer Operations*. These FMs include FM 3-34.230, *Topographic Operations*, FM 5-33, *Terrain Analysis*, and FM 34-130, *Intelligence Preparation of the Battlefield*. FM 3-34.230 covers various leader tasks and the supporting engineer topographic organizations. The latter two FMs provide specific guidance and supporting tasks for the skill of "Conduct terrain analysis."

Both the MTPs and FMs assume a foundation in a broad range of basic terrain-related SKBs. These basic tasks, or military qualification standards (MQS), are gained during officer basic training and are described in the soldier training publications (STPs).

STP 21-I-MQS, *Precommissioning Requirements*, and STP 21-1-SMCT, *Soldier's Manual of Common Tasks*, state the initial SKBs every officer should possess upon commissioning. STP 5-21II-MQS, *Engineer (21) Company Grade Officer's Manual*, covers forty-two general leader and specific branch tasks for engineer lieutenants and captains. Geospatial engineering SKBs are required in nine of the forty-two tasks. The STP recommends that these tasks be taught at the Engineer Officer Basic Course (now called the Basic Officer Leader Course (BOLC)) and Engineer Captains Career Course (ECCC). STP 21-III-MQS, *Leader Development Manual for Majors and Lieutenant Colonels*, does not provide specific tasks, as do the previous STPs, but at least four SKBs are implied for field-grade engineer officers. One challenge is that all of the STP officer manuals have not been updated since 1994 and do not account for shifts in doctrine.

Numerous historical sources demonstrate that geospatial engineering is an enduring combat capability of engineers. Captain Hotchkiss (Civil War) and Major Daniel Kennedy (World War II) both served as terrain experts for general officers. Their autobiographies, *Make Me a Map of the Valley* and *Surveying the Century,* are primary sources of note for today's terrain experts. The books *From the Golden Gate to Mexico City: The U.S. Army Topographical Engineers in the Mexican War, 1846-1848* by Adrian Traas and *Vanguard of Expansion--Army Engineers in the Trans-Mississippi West, 1819-1879* by Frank Schubert provide comprehensive coverage of the Corps of Engineers during the nineteenth century.

Operational Environment

The operational environment is the current and expected complex environment in which military operations are conducted. Threat forces, politics, combat operations, information, and technology all influence it. For this study, both the Interim Force and Objective Force are part of the operational environment.

There are past writings that discuss the challenges of a non-linear and non-contiguous battlefield. In *The Defence of Duffer's Drift*, E. D. Swinton captures twenty-two timeless tactical lessons of the British involvement in the Boer War in South Africa, 1899-1902. Several of these lessons specifically apply to today's engineer lieutenants who aid maneuver teams and task forces in small-unit combat operations.

During interviews, senior engineer leaders emphasized salient tasks that engineer officers must master for future tactical operations. Lieutenant Colonel William Goetz, commander of the 29th Engineer Battalion (Topographic), provides direct support (DS) and general support to tactical, operational, and strategic organizations throughout the Pacific Ocean region. His experiences as both a combat and topographic engineer officer provide a balanced perspective to the essential SKBs for officers. Chief Warrant Officer 5 (CW5) Ken Tatro also commented on the key tasks officers must master. He served as a terrain analysis technician over the past fifteen years, to include providing direct support to tactical commanders and engineer officers.

FM 3-34.221, *Engineer Operations -Stryker Brigade Combat Team (SBCT)*, is a recent publication that addresses how engineers operate in the Interim Force. The term SBCT is a recent change from the Interim Brigade Combat Team (IBCT). While the SBCT sharply reduced organic combat engineering capability (an engineer battalion to a company), geospatial engineering support dramatically increased with the addition of a five-soldier terrain team to the maneuver support cell embedded in the SBCT's headquarters company. This cell is staffed with an engineer major, captain, and master sergeant. At least twenty-one geospatial SKBs can be derived from this FM for the officers in the maneuver support cell and engineer company.

The Objective Force is built around brigade-size fighting elements called UAs and enhanced with additional capabilities from division and corps-level organizations called UEs. This force will dominate across the full-spectrum of military operations with the goal to "see first, understand first, act first, and finish decisively." To achieve this dominance, the Objective Force must have unprecedented mobility and information of the battlefield:

> Objective Force units will possess superior tactical mobility. Platforms will negotiate all surfaces, road, off-road, trails, water crossings, and narrow gaps. They will possess superior capability to detect presence and disposition of mines and booby traps and possess an in-stride mark and breach capability. Mounted units require the capability to conduct route reconnaissance with forward looking and off-road sensors to clear at greatly increased speeds (50+ kph). (US Department of the Army 2002e, 12)

Engineers are responsible for enabling this superior tactical mobility by providing assured mobility through development of the COP, by providing BTRA, and by conducting mechanical mobility support missions for the UA and UE (FM 3-4.221 2002, 1-2). In terms of geospatial information dominance, engineers must deliver terrain expertise and data for virtually any environment and all missions shifting back and forth across the full-spectrum of military operations. The volume, diversity, and complexity of the essential geospatial information will exceed that of current paper map and imagery based products used today.

In response to the Objective Force requirements, USAES focused its 2002 annual senior leader conference on key engineer tasks that support the full-spectrum operations of the Objective Force. The April 2002 *Engineer* magazine contains several articles on how USAES is prioritizing its efforts on assured mobility, geospatial information, and BTRA. The school also has the responsibility in the Training and Doctrine Command (TRADOC) to serve as the GI&S integrator for all Army systems and training, which it does through the TRADOC Program Integration Office--Terrain Data (TPIO-TD). Lieutenant Colonel Steve Tupper and Dr. Merrill Stevens provided information on current developments in relevant Objective Force doctrine, organization, and leadership.

Directed Skills

Directed skills are the task given to a unit or individuals other than the given wartime missions. These tasks often fall within the realm of stability and support operations (SASO), such as humanitarian assistance, disaster relief, and environmental restoration.

Directed skills often appear in the recommendations and experiences of engineer officers from around the branch in the *Engineer* magazine published quarterly by USAES. At least forty-two SKBs are directly or indirectly stated in articles between 1989 (end of Cold War) to present. Former USAES commandants, such as Major General Anders Aadland, Lieutenant General Robert Flowers, Major General Clair Gill, and Lieutenant General Joe Ballard, shared their vision of what officers must know and prepare for to keep the branch relevant in geospatial engineering. Senior civilian faculty of USAES, such as Mr. Ralph Erwin (TPIO-TD), Mr. Mike Fowler (Directorate of Combat Developments), Mr. Vern Lowery (Maneuver Support Battle Lab), Mr. Jeb Stewart (Directorate of Combat Developments), and Mr. Brian Murphy (Terrain Visualization Center), discussed doctrinal, organizational, and material changes in the branch. The members of the TPIO-TD office at USAES, such as Brigadier General

Edwin Arnold, Colonel Robert Kirby, Colonel William Pierce, Lieutenant Colonel Earl Hooper, Dr. Merrill Stevens, and Mr. Mark Adams, contributed articles describing the current and future geospatial engineering roles of officers and the Regiment. Officers with little to no prior topographic experience, such as Lieutenant Colonel C. Kevin Williams, Major Kenneth Crawford, Major David Treleaven, Major Douglas Victor, and Captain John DeJarnette, expressed the importance of knowing geospatial engineering tasks in tactical operations. Other officers with topographic backgrounds, such as Major Dirk Plante, Major Chris Kramer, and Captain Russ Kirby, exposed various terrain assets that combat engineers could exploit. Engineer terrain warrant officers, such as CW3 Terri Metzger and CW2 Chris Morken, provided their thoughts as terrain analysts for engineer officers to learn from. Some engineer branch team members, such as Dr. James Dunn, Mr. Richard Chaney, Major William Bayles, and Major Mark Adkins captured the lessons of past engineer officers in the post Civil War period, World War II and Grenada.

External Guidance

External guidance for unit METL development usually comes from MTPs and the AUTL. For engineer officers, external guidance includes Force Integration Plans for the Objective Force and recommendations from other-than-engineer sources. One historic example of such recommendations comes from a nineteenth century infantry officer. Captain Eben Swift was not an engineer, topographer, or general, but he made several observations on the importance of terrain-related tasks that are noteworthy today. In his 1897 work, "The Lyceum at Fort Agawam," he outlines a twenty-four week professional development program for the officers of an infantry regiment stationed on the western frontier. The very first requirement of the Lyceum was for the junior officers to compile a tactical map of the thirty-two square mile training area on which they would continually improve upon throughout the course of instruction. In World War II, several other non-engineer military officers, such as Norman Maclean and Harry Musham, developed substantial training manuals for military topography to prepare officers and soldiers heading off to war.

External guidance also comes from interviews with officers at Fort Leonard Wood, Missouri, and Fort Leavenworth, Kansas. The interviews involved three similar sets of questions dealing with terrain expertise: one for future battalion and brigade commanders at Fort Leavenworth's Pre-Command Course (PCC); one for engineer majors attending the Command and General Staff Course (CGSC); and one for engineer captains at ECCC. These interviews provided primary source information on what engineer officers should focus and train on. The interview questions and raw feedback are in Appendix B.

What is an Engineer officer terrain expert?

One of the first discussions of engineer officers being terrain experts came from the then Colonel Edward Arnold, interim Director of TPIO-TD, at Fort Leonard Wood in 1997. He coined the label terrain visualization expert in the article "Being a Terrain Visualization Expert." He described the importance of terrain visualization for the maneuver commander and how engineer officers and terrain warrant officers must work together to accomplish this. He states that terrain appreciation and terrain evaluation are skills that should be second nature to any engineer officer (1997, 22).

The past five USAES commandants have also promoted that engineers are the Army's terrain experts. Major General Anders Aadland stated that:

> Geospatial engineering is the development, dissemination, and analysis of terrain information that is accurately referenced to precise locations on the earth's surface. Although this is new terminology (replacing topography), the emphasis is still on engineers being the terrain experts for the maneuver commander. . . . Engineer leaders must know how to exploit this information. (Aadland and Allen 2002, 8)

Army field manuals, as a whole, only mention engineer officers as terrain experts in passing as shown in the following excerpts:

> FM 5-10, *Combat Engineer Platoon*, simply states, "the platoon leader must advise the maneuver commander on the military aspects of the terrain since he is the terrain expert." (1995, 2-1)

> FM 5-100, *Engineer Operations*, states, "The engineer is the terrain expert. He must work closely with the S2 to determine advantages and disadvantages the terrain gives the attacking force." (1996, 8-3)

> FM 17-95, *Cavalry Operations*, states, "The regimental engineer is the terrain expert." (1996, 2-1)

> FM 90-13, *River Crossing Operations*, states, "Engineers analyze the terrain to determine the maneuver potential, ways to reduce natural and enemy obstacles, and how they can deny freedom of maneuver to the enemy by enhancing the inherent obstacle value of the terrain. . . . The engineer is the terrain expert." (2000, 2-1)

> FM 3-34.230, *Topographic Operations*, describes the Engineer officer's geospatial role as the terrain visualization expert. This expert assists the commander in visualizing the terrain, identifying and understanding terrain aspects for exploitation by friendly and enemy forces, and providing subjective evaluation of the terrain's physical attributes and physical capabilities of vehicles, equipment and people. (2000, 3-1)

Though none of these references provide a clear definition, they all indicate several important tasks for the terrain expert. First, the engineer officer must be able to connect geospatial engineering with tactical operations. He is often referred to as an advisor or assistant to the tactical commander for exploiting the terrain. Second, the terrain expert must be able to generate, obtain and/or use geospatial products, such as overlays. Third, he should have a close working relationship with the military intelligence staff and the engineer terrain warrant officer, the terrain analysis and GI&S expert. These three tasks lead to the following recommended definition.

<u>Terrain Expert</u>. One who demonstrates skills, knowledge, and behaviors in rendering geospatial engineering to the tactical situation in order to take advantage of the battlespace environment. The expert understands the limits and capabilities of GI&S and can integrate them into the appropriate tactical language and processes. Engineer doctrine states that the engineer officer is the terrain expert (Tupper 2003).

In a sense, the engineer officer serves as a bridge between the complex geospatial engineering sciences and services performed by terrain analysis technicians and topographic analysts and the highly fluid needs of tactical operations performed by the maneuver and fire support BOSs. He must deliver the right terrain expertise for each tactical situation. Thus, he does not need to personally know how to accomplish every geospatial task in detail, just as he does not need to know how to operate every weapon and communication system of the combined arms team. An engineer officer must master the SKBs that bring the two functions together. Major General Laporte highlights this important connection in his article "Terrain Visualization." He concludes that after the division terrain team studied the terrain properties and weather effects on the ground, the "combat engineers interpreted and evaluated the products and data to produce the estimates for mobility, countermobility, and impact upon every Battlefield Operating System during the fight" (Laporte and Melcher 1997, 76). Colonel Melcher, the engineer brigade commander, and his engineer officers were the 1st Cavalry Division's terrain experts.

Existing and Potential Skills, Knowledge, and Behaviors

In this section, SKBs are selected from requirements in the five sources of inputs. Many of these SKBs support one another, as well as other engineer missions and functions. Terrain expertise is a foundational element in providing other engineer services, since virtually everything an engineer must do involves the ground. Some of these tasks are branch immaterial--they apply to officers in their role as soldiers and leaders. Engineer officers performed other tasks during distinct periods, but not today. In all cases, the proposed task is underlined and described by the literary or field sources that support it.

Geospatial Function #1: Terrain Analysis

Provide input to IPB. This is a specified behavior for all engineer captains in STP 5-21II-MQS task number O1-2250.20-1006. Engineers support IPB in four important ways: they evaluate existing GI&S databases to identify gaps, assist in the development of intelligence requirements for collection, perform terrain analysis, and describe the enemy and friendly engineer capabilities (FM 34-130 1994, 2-3). FM 5-100 states the engineer "analyzes the terrain and weather and assesses the impact that they will have on military/engineer operations. He analyzes the terrain using the following five military aspects of terrain (OCOKA)" (1996, 7-1). In a report focused on IPB, the observer controllers (OCs) at NTC remarked that "IPB is not just the responsibility of the S2. . . . Each staff officer should analyze their specific BOS and provide that analysis to the S2" (*CTC Newsletter* 96-12 1996, Foreword). By the rank of major, engineer officers are expected to be experts in the conduct of IPB (STP 21-III-MQS 1993, 10).

Conduct terrain analysis using OCOKA. This skill is a precommissioning requirement found in STP 21-I-MQS, task number 04-3306.01-0008. This skill is integral to many tactical processes, such as IPB and MDMP, as well as several engineer unit collective tasks, such as Prepare an Engineer Estimate (05-2-0002.05-R01D), Perform Engineer Battlefield Assessment (05-2-0027.05-R01D), and Analyze Battlefield Information (05-2-0415.05-R01D). As early as 400 B.C., Sun-Tzu's emphasized the need to analyze the terrain, "These six are the principles connected with Earth. The general who has attained a responsible post must be careful to study them"(Griffith 1963, 129). In a lesson from the Boer War, E.D. Swinton specifically notes the importance of identifying the dead-space created by relief and convex ground that masks observation of enemy and friendly maneuver (1986, 58). Today, a number of FMs include terrain analysis as an essential task. FM 3-0, *Operations*, explains that terrain analysis must not only include OCOKA analysis of natural terrain elements, but also account for man-made features, the impact of weather, environmental contamination, and three dimensional complex topography of urbanized terrain (2001, 5-15). FM 5-100 states that the function of terrain analysis is "to reduce the uncertainties regarding the effects of natural and man-made terrain on friendly and enemy operations" and that assists "the commander in establishing the proper tempo of the offense" (1996, 7-1 and 8-3). FM 34-130, *Intelligence Preparation of the Battlefield*, and FM 5-33, *Terrain Analysis,* describe in detail how to conduct terrain analysis using OCOKA and apply its effects on military operations. The conduct of terrain analysis is often covered in the CALL publications that review unit performance at the CTCs.

Identify terrain features on a map. This knowledge is a common soldier task in STP 21-I-SMCT, task number 071-329-1001.

Prepare a Modified Combined Obstacle Overlay (MCOO). FM 5-100 describes this as a task that all engineer S2s perform during tactical planning. The MCOO is a graphical terrain analysis on which all other IPB products are based (1996, 7-1). The OCs at JRTC remind units that "Displaying a doctrinally correct MCOO is not enough;" it must pass the "so what" test (*CTC Trends* 99-7 1999, 7).

Identify weather impacts on the terrain. This knowledge supports the unit collective task of "Prepare an Engineer Estimate" (ARTEP 5-335-65-MTP 2000, 5-166).

Validate seasonal terrain feature data, such as vegetation and hydrology conditions. Maps and imagery may either be outdated or out-of-season for when operations will occur. Seasonal features should be verified with ground or aerial reconnaissance (Crawford 1998, 42). In his study of twenty-seven battles on four different continents, Harold Winters observed that the species, size, density, structure and distribution of vegetation can greatly influence tactical operations, such as at the Battle of the Wilderness in 1863 and in the Ia Drang Valley of Vietnam in 1965. Furthermore, the "role of terrain, weather, climate, soil or vegetation in one battle is by no means a reliable predictor of its effect on the next [battle]" (1998, 3 and 111).

Understand the military geography for different regions. CW2 Metzger discussed the unique challenges of analyzing the military geography of Saudi Arabia and Kuwait in the immediate weeks after Iraq's invasion of Kuwait in August 1990. The desert terrain involved significantly different topographic symbols and imagery interpretation than had been practiced for the entrenched European scenarios (1992, 26). Today, the current

Chief of Engineers, Lieutenant General Flowers, makes it a frequent practice, as a senior engineer leader, to quiz fellow engineers on their knowledge of military geography in areas of global and US interest. He stresses that leaders should maintain situational awareness of events and locations that could lead to military actions.

Use terrain to deceive the enemy. E. D. Swinton points out that any good analysis of the terrain should identify how terrain features, weather, and illumination can be used against the enemy's likely actions and weaknesses (1986, 58).

Conduct terrain analysis for urban environments. FM 3-0 stresses the challenge urban areas pose to today's military operations and forces. The manual encourages commanders to "view cities not just as a topographic feature but as dynamic entities. . . . Planning for urban operations requires careful IPB, with particular emphasis on the three-dimensional nature of the topography and the intricate social structure of the population" (2001, 6-77). The SBCT engineer officer must request, manage, and analyze urban area geospatial products for the brigade (FM 3-34.221 2002, 9-7). In 1991, Major Kevin Johnson prepared the monograph "Intelligence Preparation of the Urban Battlefield," and in 2000, Major Willard Burleson, III, wrote the thesis "Mission Analysis During Future Military Operations in Urbanized Terrain (MOUT)." Both studies describe the challenges of making the doctrinal terrain analysis process fit the frequent operations in urbanized areas around the globe. In the article, "Terrain Analysis Considerations," Major Chris Kramer at USAES adds new challenges and solutions to terrain analysis caused by the exponentially growing number of digital terrain databases. The OCs at the CTCs reiterate that urban terrain analysis is substantially more difficult and too complex for the standard terrain analysis most units conduct (*CTC Newsletter* 99-16 1999, 1-6).

Conduct terrain analysis for peacekeeping operations. While assigned as the G2 Geographic Officer on a brigade-sized United Nations task force in 1995, the then Captain David Treleaven noted that traditional terrain analysis using OCOKA was inadequate for addressing the terrain's impact on military operations other than war (MOOTW). This was especially true when the enemy is not a known or standardized force (1995, 10).

Prepare engineer estimates to include geospatial engineering capabilities. This specified skill is for all engineer captains from STP 5-21II-MQS, task number O1-2250.20-1001, to be trained at ECCC. The primary engineer estimate conducted during tactical decision making process at all levels is the EBA. The OCs at NTC noted that engineer company executive officers as task force planners, usually first lieutenants, did not conduct EBA to standard in the first and second quarters of FY01 (*CTC Trends* 02-17 2002, 16, 22). During EBA, the engineer officer must prepare an estimate of friendly and enemy engineer unit mobility, countermobility, survivability, and general engineering capabilities to understand what is available to physically shape the terrain. This should also include the enemy's capability to alter the landscape. In the months following Iraq's invasion of Kuwait, the Iraqi army quickly added extensive minefields, obstacle belts, and new and improved roads throughout Kuwait, requiring constant re-evaluation by the ISR assets in theater (Metzger 1992, 27). After the Gulf War ended, Iraq drained and canalized the lower Tigris River and Euphrates River, eliminating hundreds of square miles of swamp while creating hundreds of linear miles of levees.

Understand the capabilities and effects of enemy weapons. Part of IPB is the assessment of enemy weapons systems. Engineers need this information to evaluate the effectiveness of terrain features, such as vegetation and masking, and protective materials to counter these weapons. E. D. Swinton repeatedly points out the challenge of integrating individual and unit survivability, firepower, and concealment against the enemy's weapons in the Boer War (1986, 35).

<u>Understand how terrain affects unit camouflage operations</u>. This knowledge supports the unit collective task of "Camouflage Vehicles and Equipment" (ARTEP 5-335-65-MTP 2000, 5-107). A leader must also understand the capabilities of opposing force surveillance to detect camouflage in the battlespace.

<u>Visualize the terrain</u>. In order to successfully aid maneuver commanders to visualize the terrain for combat operations, engineer officers must be able to see the terrain as the commanders and their staffs would. They also must be able to see how the terrain impacts engineer operations (Gill 1996, 13).

<u>Survey the terrain to identify the critical terrain elements for a unit defensive position</u>. This skill is required to perform the unit collective tasks "Fight as Infantry" and "Establish Company Defensive Position" (ARTEP 5-335-65-MTP 2000, 5-39, 5-51). The leader must be able to perform this skill in both urban and non-urban terrain.

<u>Evaluate terrain 360-degrees around the chosen location</u> In the Boer War, the British faced an operational environment similar to the non-linear, non-contiguous battlespace expected for the Objective Force. Since the enemy can attack from any direction, terrain analysis must not be limited to a certain "front" (Swinton 1986, 47).

<u>Supervise site selection and layout</u>. This is a specified behavior for all engineer lieutenants in STP 5-21II-MQS. It demonstrates how geospatial engineering can support general engineering missions, such as bridge and building construction.

Provide terrain analysis for Deep Operations Coordination Cell (DOCC) activities. An engineer staff officer at division or corps normally sits in on the DOCC for tactical operations. Specifically, the engineer identifies features that influence enemy mobility and recommends targets to affect this. The OCs in the Battle Command Training Program (BCTP) noted that engineer units perform inconsistently in this task (Light 1999, 58).

Geospatial Function #2: Data Collection

Evaluate the availability of standard and nonstandard map products. The first step of IPB, "Define the Battlefield Environment," requires that leaders "evaluate existing data bases and identify intelligence gaps" (FM 34-130 1994, 2-3). FM 5-100 describes this as a routine staff function at brigade through corps levels (1996, 12-10).

Request standard NIMA products through logistics channels or directly from the Defense Logistics Agency (DLA). Major Plante, the GI&S Officer for US Forces Korea, noted that every engineer officer should have this skill in order to educate the staffs and units they support. Standard products now have national stock numbers (NSNs) for units to order products either manually or on-line (1999, 38).

Obtain standard and non-standard terrain-products through controlled sources, such as the SIPRNET. This is a general research observation of existing internet capabilities. Both NIMA and Army topographic units, to include terrain teams, can post their products for limited distribution with secure access. A battalion or brigade S2 would normally require this skill.

Understand the geospatial capabilities available in joint operations. All engineer majors should have this knowledge according to STP 21-III-MQS. In FM 5-100, the chapter on contingency operations prescribes that engineer leaders should coordinate for early collection of terrain information through reconnaissance, topographic survey, and satellite imagery. They should also know what terrain analysis and topographic reproduction capabilities are available to the joint task force (1996, 12-14).

<u>Understand how space-based systems can enhance warfighting capabilities at the tactical and higher levels of war in reconnaissance, position and navigation, and weather</u>. All engineer majors are expected to have this knowledge (STP 21-III-MQS 1993, 13).

<u>Understand the foundation data (FD) construct from NIMA</u>. The FD is a GI&S construct for data that officers must know and learn how to exploit (Aadland and Allen 2002, 8). It is the baseline for geospatial information at roughly 1:250,000 scale. It will provide near global coverage of five- meter resolution imagery, elevation data, and feature foundation data (FFD). The primary FD product for tactical levels is the mission specific data set (MSDS). The MSDS is a tailored set of products at approximately 1:50,000 scale designed to meet a maneuver commander's data requirements (Pierce 2001, 10).

<u>Understand map datums and scales</u>. Major Plante points out that all engineer officers need to be able to articulate what and how map datums can impact GI&S products and global positioning system (GPS) equipment (1999, 39).

<u>Understand "Reachback" capabilities for GI&S support in the SBCT and UA</u>. The SBCT has an organic terrain detachment to provide immediate GI&S support, but it requires GI&S updates from in-theater and home-station databases. SBCT engineer officers must be familiar with the capabilities and limitations of the terrain detachment in transmitting GI&S data (FM 3-34.221 2002, 1-7).

Conduct a reconnaissance. This skill is a precommissioning requirement found in STP 21-I-MQS, task number 04-3302.01-0003. Both FM 5-100 and FM 5-170 discuss in detail the necessity of engineers conducting reconnaissance to gather data about the battlespace. Engineer leaders must be able to accomplish technical, tactical, and engineer forms of reconnaissance (ARTEP 5-335-65-MTP 2000, 5-12).

Submit both verbal and written reconnaissance reports as required by Standardization Agreement (STANAG) 2003. This behavior supports the unit collective task of "Conduct Report Procedures." These reports include a description of terrain, deepness of ravines and draws, bridge conditions, effect on track/wheeled vehicles, and any map corrections (ARTEP 5-335-65-MTP 2000, 5-183).

Direct engineer reconnaissance missions. This is a specified behavior for all engineer lieutenants in STP 5-21II-MQS. Both FM 5-100 and FM 5-170 discuss that reconnaissance is used to verify the accuracy of initial terrain assessments. Leaders must identify specific requirements, augment patrols and scouts to collect essential information, and integrate engineer reconnaissance into the maneuver commander's plan (FM 5-100 1996, 8-5).

Perform military sketching. Captain Swift noted that all junior officers of his time could perform military sketching in order to draw a basic map of the terrain on which military activity might be conducted. Throughout much of the nineteenth century, military sketching was taught at most of the European military academies. Until 1999, all cadets at the US Military Academy (USMA) received at least some training in topographic sketching in the required undergraduate course "EV203 Terrain Analysis" (Starke 2003). In the active and reserve component, the modification table of organization and equipment (MTOE) of every combat engineer platoon contains a military field sketching for the purpose of making field sketches, reconnaissance, minefield recording and surveying.

Record terrain information daily. Captain Swift recommended that all officers be in the daily habit, or behavior, of noting and recording the military characteristics of the terrain around them. Their notes were to be incorporated in a "progress map" maintained by a designated officer. This practice also enhanced their memory of the ground (Swift 1897, 268).

Coordinate with the S2/G2 and S3/G3 for collecting terrain information. This task is a behavior for all engineer officers in staff positions on brigade and higher staffs (FM 3-34.221 2001, 3-8).

Establish information requirements (IR) for essential elements of terrain or engineer information. This task is a specified behavior for all engineer captains in STP 5-21II-MQS that supports the unit collective tasks of "Conduct an Engineer-Intelligence Collection" and "Conduct a Tactical Reconnaissance" (ARTEP 5-335-65-MTP 2000, 5-19). The TFE and commander prepare the reconnaissance and surveillance (R&S) plan. The OCs at NTC note that staff engineers, especially the engineer battalion S2s and ABEs, failed to prepare R&S plans during rotations in 1998 and 2001 (*CTC Trends* 99-10 1999, 47; *CTC Trends* 02-17 2002).

Track templated and known obstacles (friendly and enemy). This task normally supports mobility and countermobility missions. Obstacle intelligence is normally templated on a graphical--analog or digital--overlay (FM 3-34.221 2001, 2-13).

Geospatial Function #3: Data Generation

Coordinate with the S2/G2 to define, prioritize, and request topographic products. All field-grade engineer officers at brigade, division and corps should be able to perform this task (FM 3-34.230 2000, 3-8).

Understand the organization and capabilities of the DS corps topographic engineer company. This is general knowledge that all engineer captains and above should have, especially when working with division and corps level staffs (FM 5-100 1996, 2-15).

Prioritize and task the production of the DS Corps Topographic Engineer Company. The senior engineer in the corps or his designated engineer staff officer performs this task to assist the corps commander and staff (FM 3-34.230 2001, 3-8).

Transmit essential terrain information to terrain teams for product update. FM 5-100 states, "As the terrain is modified (bridges destroyed, roads built), the terrain team updates its data base and issues new products. Necessary information is reported through engineer channels" (1996, 9-4).

Provide the status of infrastructure for contingency operations. A captain or higher should provide this information to a brigade or higher commander or staff (FM 5-100 1996, 12-4).

Execute target-folder battle drills. This unit collective task is normally executed by company-grade engineer officers (ARTEP 5-335-65-MTP 2000, 5-23).

Geospatial Function #4: Database Management

Establish data and database management practices. This is a responsibility of the senior engineer officer on the maneuver staff (Erwin 2001, 16).

Disseminate terrain analysis and other geospatial products. The OCs at NTC advise that engineer units must prepare terrain products as soon as they find out the area of operations (AOs) and area of interests (AIs). Once deployed, there often is not adequate time to generate products that can be distributed to the (Bell 1999, 41).

Resolve differences between various reports and products to render a single COP. Engineer officers, utilizing the supporting terrain team, must de-conflict contrary terrain data covering the same ground (Hooper, Morken, and Murphy 2001, 15). Mr. Erwin further emphasizes that the senior engineer officer and terrain warrants must ensure all automated systems are operating on the "same sheet of music" as far as a common topographic data set (2001, 16). Major Plante emphasized that engineer officers should also inform the warfighter of changes that occur both universally to geospatial products, such as datum changes (1999, 40).

Integrate nonstandard and non-US GI&S products into tactical databases. This is a task all engineer captains should know. Captain Treleaven relates that local maps in Bosnia were of better quality than available maps from NIMA, but had to be "regridded" (1995, 10). French forces in World War II and US forces in the invasion of Grenada had to rely on commercial road maps to orient activities.

Maintain and update the map unit basic load (UBL) for a company or battalion. Each unit may have a map account and basic allocation of products that NIMA will automatically provide. Most units, to include engineer companies and battalions, are unaware of this opportunity and necessity (Plante 1999, 39).

Geospatial Function #5: Data Manipulation and Exploitation

Use PC-based terrain analysis tools, such as TerraBase II, to create tactical decision aids (TDAs). Personal computer (PC) based tools include TerraBase II, ArcView, and Falcon View. FM 5-71-3, *Brigade Engineer Combat Operations (Armored)*, dedicates Annex C to the tactical application of TerraBase II. In 2001, the OCs at NTC noted that engineer units--especially TF and brigade engineer staffs--are often weak or lacking in using terrain analysis tools to improve planning and terrain visualization (*CTC Trends* 02-17 2002). Several writers to the *Engineer* explain how TerraBase II provides a simple and effective tool for the engineer officer. Articles include "TerraBase II, Version 3.0--Supporting the Terrain Visualization Expert" (Hooper and Adams 1998, 30), "Introducing TerraBase II" (Kirby 1997, 38), and "Engineer Support to Engagement Area Development" (Crawford 1998, 41). Engineer leaders in the reserve component seldom get support from a terrain team or DTSS. Lieutenant Colonel Rensema, commander of the 164th Engineer Battalion in Minot, North Dakota, relates how his staff successfully combined the effects of TerraBase II and ArcView to plan a potential river crossing operation (Rensema, Erickson, and Herda 2000, 34). The three officer groups interviewed indicated that training and use of TerraBase II is absolutely essential for engineer officers, especially in company-grade positions.

Understand the BTRA capabilities embedded in ABCS platforms. Engineer leaders should be knowledgeable in the terrain analysis and terrain visualization capabilities that exist and emerge in the ABCS platforms that will reduce manual calculations. Engineer officers should know the capabilities and limitations of these systems (Fowler and Johnston 2002, 13).

Understand the capabilities of JMTK as a component of ABCS. Engineer leaders should be familiar with JMTK's components to render GI&S for tactical applications (Snyman 2002, 19). Commercial JMTK (C/JMTK) is the future replacement of JMTK for the Objective Force.

Perform grid coordinate conversions. Engineer officers should understand how to convert grid coordinates for differing map datums (Treleaven 1995, 10). Coordinate conversions can be accomplished manually or using various software programs.

Use a digital situational awareness (SA) overlay to conduct a map reconnaissance The SA overlay is on the FBCB2 system and MCS (ARTEP 5-335-65-MTP 2000, 5-9). This leader skill supports the accomplishment of four engineer company collective tasks: "Conduct a Tactical Reconnaissance," "Conduct a Water-Crossing and Site-Approach Reconnaissance," "Control Combat Formations," and "Conduct a Radiological or Chemical/Biological Reconnaissance or Survey." The OCs at NTC commented that engineer platoon leaders and platoon sergeants "must train on route selection based on map reconnaissance and the leader's reconnaissance to ensure efficient and safe travel of heavy equipment in rough terrain." (Prude 1999, 35)

Use a digital SA overlay to conduct a map orientation. The SA overlay is on the FBCB2 system and the digital reconnaissance system (ARTEP 5-335-65-MTP 2000, 5-12). This leader skill supports the accomplishment of "Conduct an Engineer Reconnaissance."

Understand the capabilities of an Engineer terrain team and the DTSS. Maneuver commanders and terrain analysis warrants officers expect engineer officers to understand the capabilities and limitations of a terrain team (CW5 Tatro 2002). In the PCC interviews, senior leaders repeatedly stated that engineer officers must know, use, and advertise the capabilities of the terrain teams. Engineer company commanders are specifically challenged to have knowledge of the DTSS and its products in order to prepare obstacle plans using the MCS (ARTEP 5-335-65 2000, 5-95). In 1996, OCs at JRTC noted that terrain teams greatly enhance a brigade or TF terrain analysis with satellite imagery products (*CTC Trends* 96-9 1996, II-1). This positive trend continued in 1997 at JRTC where OCs reported that "S2 sections and their supporting topographic teams are preparing detailed terrain analysis products" (*CTC Trends* 97-19 1997, 1). But in 1999, the OCs observed that not all commanders and their staffs were familiar with the terrain team's capabilities and limitations nor did they routinely request a package of standard terrain analysis products (*CTC Trends* 99-7 1999, 3).

Understand the different types of digital GI&S data and their uses. Terrain analysis warrant officers expect engineer officers to be familiar with the common types of digital data in order to translate the commanders' and staffs' requirements into tactical decision aids (CW5 Tatro 2002). GI&S data comes in a variety of formats (raster, vector, matrix, and text data), types (imagery, maps, and elevation data), resolutions, and scales. Major Plante noted that every engineer officer should be very familiar with the various GI&S products in order to educate the staffs and units they support. Most units are only familiar with paper 1:50,000 and 1:250,000 map products (1999, 38).

Understand the digital size of standard NIMA and DTSS GI&S products for ABCS use. Terrain analysis warrant officers expect engineer officers to know the impact that DTSS products make on the ABCS architecture (CW5 Tatro 2002).

Geospatial Function #6: Cartographic Production and Reproduction

Manage printing and survey assistance for the rapid replication of topographic products. The senior corps engineer officer is responsible for this behavior (FM 3-34.230 2000, 3-9).

Prioritize terrain product production. Engineer officers must understand the time it takes to acquire and produce products so that terrain teams are not overwhelmed with unnecessary projects (Chaney 1998, 16).

Provide bridge classification maps. This is a specific task all engineer officers should be familiar with (FM 5-100 1996, 9-4). The OCs at JRTC state that engineers should know how to provide bridge and road classifications to the S2 and brief their limitations to the commander (*CTC Newsletter* 98-10 1998, 1-7).

Geospatial Function #7: Geodetic Survey

Understand the role of the survey platoon in the DS corps topographic engineer company. While few engineer officers direct survey operations, engineer officers at division staff levels and higher should understand this geospatial engineer function.

Understand the levels of accuracy provided by geodetic survey and GPS. Similar to the task to understand map datums and scales, engineer officers should understand the capabilities and dangers of commonly used GPS equipment as compared to the level of accuracy provided by a survey platoon from the DS corps topographic engineer company.

Providing Terrain Advice

<u>Understand and coordinate the seven geospatial engineering functions at tactical echelons</u>. All engineer lieutenants should be able to brief engineer battlefield functions, to include the appropriate geospatial engineering capabilities available to their respective maneuver commanders. The EBA is one means of sharing this information. All company-grade officers need this knowledge to accomplish the unit collective task "Identify topographic support requirements" from ARTEP 5-335-65-MTP. All senior engineer officers at the brigade, division, and corps levels also should have this knowledge (FM 3-34.230 2000, 3-8).

<u>Understand the roles of the engineer officer, terrain analysis warrant officer and topographic analyst</u>. This is an implied knowledge for all engineer officers in order for them to fully exploit and maximize the expertise available to the commander and staff (Tatro 2002).

<u>Prepare the topographic annex or appendix to tactical plans and orders</u>. This is a specific behavior for the senior engineer officer in the brigade, division, or corps per FM 3-34.230 and FM 3-34.221.

<u>Integrate a terrain team in brigade operations</u>. Every engineer captain and higher should have this knowledge (FM 3-34.221 2002, 1-10). The officers of the 3rd Brigade Combat Team (BCT) and 4th Engineer Battalion pioneered this integration during the 1997 Division Army Warfighter Experiment. They recommended that the terrain team merge early with the BCT early in the planning process (pre-deployment), and if possible, establish a habitual relationship with the BCT (Chamberlain, Williams, and

Perez 1998, 21). The SBCT establishes this permanent relationship between terrain team and BCT.

Identify critical logistics requirements for organic geospatial engineering support. Virtually every war starts with both sides severely lacking maps. Even during wars, terrain products can remain critically short. After the Allied invasion of Normandy, the US forces quickly ran out of maps due to a paper shortage. French paper was inadequate, so over 10 million maps were printed on the reverse side of captured German maps. Transportation assets to distribute maps also ran critically short throughout the remainder of the war (Chaney 1998, 17). Logistical challenges can also include the deployment of unique topographic equipment early in the time-phased force deployment list (TPFDL) in order to provide geospatial engineering support for forces as they arrive (Wright 1992, 3). Engineer officers must have knowledge of the critical logistical requirements for corps and below topographic assets.

Serve as the corps topographic officer. The senior engineer officer in the corps or the chief of the staff engineer section is the corps commander's geospatial expert (FM 3-34.230 2000, 3-9).

Identify threats/risks to geospatial operations and functions. Not only were US forces plagued with paper shortages in World War II, but enemy action also denied the Allies from having adequate maps in 1944. Most of the advance maps sank in the harbor at Cherbourg during the first days of the Normandy invasion. Late in June, storms caused heavy losses of printing equipment and supplies (Chaney 1998, 16). Engineers must still be aware of the threats to geospatial engineering from enemy (cyber, infrastructure, command and control (C2), and transportation system attacks) and weather (cold, heat, dust, and humidity).

Advise the commander on the use of terrain for combat operations. This task is a specified behavior for all engineer captains in STP 5-21II-MQS to be trained at ECCC. FM 5-100 states "the engineer must be able to advise the maneuver commander on the advantages and disadvantages of each piece of terrain from the friendly and enemy's points of view" (1996, 9-4). ARTEP 5-335-65-MTP lists this behavior as a primary supporting task for the unit collective task "Conduct Breaching Operations." The OCs at NTC recommended the commander receive terrain products that show key and decisive terrain along with their significance (*CTC Newsletter* 96-12 1996, III-2). A year later, the OCs at NTC made a more disturbing observation on engineer officer's inability to advise the commander. Even though the ABE was briefing terrain analysis during MDMP and OPORD briefs, the OCs recommended that the S2 brief the terrain and its significance instead the ABE (*CTC Trends* 97-16 1997, 9). During 2000, the OCs at JRTC noted that staff officers, especially company grade staff officers at TF and brigade were not trained to be staff officers. While they are well trained to be leaders they generally lack the technical foundation to enhance the staff (*CTC Trends* 01-2 2001, 38). Lieutenant Colonel Rensema and Captain Herda of the the 164th Engineer Battalion emphasize that engineer officers must advise the commander on the effects of terrain, regardless of component and terrain team support (Rensema, Erickson, and Herda 2000, 34).

Assist the maneuver commander with terrain visualization. While terrain visualization may in part rely on the commander's imagination, it should not fail due to lack of geospatial information. TRADOC states this responsibility as:

> Terrain visualization is a basic and fundamental leadership skill. A battle commander must understand how terrain influences every aspect of military operations. . . . Engineers have the responsibility to advise commanders on the effective use of the terrain. (TRADOC Pam 525-41 1997, 2-1)

One means to enable better terrain visualization is through hard and soft copy products saved in a terrain visualization mission folder (FM 3-34.221 2002, 2-13).

Brief terrain effects. This is an implied task for all engineer officers based on the doctrinal references to being a terrain expert. This task is accomplished during MDMP and in planning and executing unit operations.

Understand the importance of military topography and map production to senior leaders. In the mid-1700's, Frederick the Great was first to elevate military topography to a distinct, prominent role in the Prussian army. He established a map and plans room in his castle in Potsdam and retained tight control on all map reproduction. In 1816, the topographical surveying service was placed under the Prussian General Staff. In 1777, George Washington appointed Robert Erksine as the first geographer of the Continental Army to create greatly needed maps (Prol 2002). The French, under Napoleon, also had a Topographic Bureau in the General Staff of the Imperial Headquarters that maintained over 500 different maps of Prussia. By 1862, the Russian General Staff maintained a Topographical Corps of 450 officers and men to manage its extensive mapping program (Wahlde 1960, 8). After the Civil War, US Army topographic engineer officers were reassigned to division staffs specifically to prepare maps for field operations (Dunn 2002, 53).

Translate GI&S into tactical terms for warfighters. As the USAES commandant, Major General Gill exhorted engineer officers to "speak the warfighter's language and avoid bombarding him with highly technical language, endless 'topospeak,' and unique acronyms" (1996, 13).

Translate tactical requirements into GI&S products and analysis. Maneuver commanders should expect their supporting engineer officers to request, develop, and provide the right TDAs for the tactical mission (Gill 1996, 15).

Understand geospatial engineering's role in assured mobility. In order for the maneuver units to have assured mobility, engineers must seek to achieve geospatial information dominance (FM 3-34.221 2002, 1-13).

Prepare a risk assessment when the lack of terrain information creates uncertainty. Maneuver commanders should expect this behavior from their supporting engineer officers (Gill 1996, 15).

Understand geospatial information requirements of each BOS. All engineer majors are expected to have this knowledge (STP 21-III-MQS 1993, 10). FM 34-130 dedicates an entire chapter to the essential terrain factors for each BOS element. Major General Laporte expected his divisional engineers to analyze the terrain and weather impact on each BOS (Laporte and Melcher 1997, 76).

Provide geospatial engineering advice to S4/G4 for main supply route (MSR) & logistics operations. The engineer major and captain in the SBCT maneuver support cell are specifically tasked to perform this behavior (FM 3-34.221 2002, 2-13).

Understand what topographic support is available for MOOTW. The engineer major and captain in the SBCT maneuver support cell are specifically tasked to have this knowledge (FM 3-34.221 2002, 6-46).

Navigate Using Geospatial Information

Perform map reading. This task is a basic skill for all soldiers in STP 21-I-SMCT.

Determine the grid coordinates of a point on a military map. This task is a basic skill for all soldiers in STP 21-I-SMCT, task number 071-329-1002.

Measure distance on a map. This task is a basic skill for all soldiers in STP 21-I-SMCT, task number 071-329-1008.

Orient a map to the ground by map terrain association. This task is a basic skill for all soldiers in STP 21-I-SMCT, task number 071-329-1012. In 1897, Captain Swift felt that officers should not only be able to orient a map to the ground but also understand the differences and relationships between the map and the ground. He warns that a map can only give but a poor picture of the terrain, and its defects are quickly detected when exercising out on the ground itself. He asserts that the best map is one where "the real ground [is] drawn in full day upon the human retina" (1897, 268).

Identify topographic symbols on a military map. This task is basic knowledge for all soldiers in STP 21-I-SMCT, task number 071-329-1000. Norman Maclean and Everett Olson assert, "Any mistake in operations due to failure to read a map correctly is absolutely inexcusable. Officers must be proficient in military topography" (Maclean and Olsen 1943, 8). Today, FM 3-25.26 (former FM 21-26), *Map Reading and Land Navigation,* and FM 21-31, *Topographic Symbols,* cover this subject in detail. FM 21-31 has not been updated since 1968 and does not account for changes in topographic symbols used in digital terrain data and products.

Determine a location on the ground by terrain association. This task is a basic skill for all soldiers in STP 21-I-SMCT, task number 071-329-1005. The OCs at NTC noted a consistent weakness in all units for their inability to conduct land navigation and to appreciate the effects of terrain. They caution units to not overly rely on GPS since "batteries fail and equipment breaks!" (Prude 1999, 35)

Determine a magnetic azimuth using a lensatic compass. This task is a basic skill for all soldiers in STP 21-I-SMCT, task number 071-329-1003.

Navigate using a map and a compass. This skill is a precommissioning requirement found in STP 21-I-MQS, task number 04-3303.01-0034.

Determine direction without a compass. This task is a basic skill for all soldiers in STP 21-I-SMCT, task number 071-329-1018.

Navigate using GPS equipment. This task is not officially coined in any of doctrinal or training manual, yet it is widely practiced throughout the Army with even greater use expected in the years to come. The GPS is replacing the compass as the primary navigational aid. GPS is an integral component of Interim Force and Objective Force C4ISR systems.

Summary

This chapter investigated a wide variety of literary and field sources to define the term terrain expert and extract eighty-seven potential SKBs. The broad parameters of ART 1.1.1.5 "Conduct Geospatial Engineering" allowed for the collection of a wide range of tasks, based on Army tactical requirements. Some SKBs are geospatially focused, such as understanding the capabilities of topographic companies and terrain teams. Other SKBs reflect how fundamental GI&S is in engineer leader and unit tasks, such as in reconnaissance missions and land navigation. Overall, the research reinforces that the tactical levels of the Army require terrain expertise from engineer officers.

CHAPTER 3

RESEARCH METHODOLOGY

Do essential things first . . . Nonessentials should not take up time required for essentials. (FM 25-100 1988, 2-1)
General Bruce C. Clarke

Introduction

This chapter evaluates the eighty-seven officer SKBs identified in the previous chapter. To assist in selecting the SKBs, five criteria are set forth. Each criterion requires a "yes" or "no" subjective evaluation that will assist in an objective scoring to determine the right SKBs to be a terrain expert. These criteria were chosen because they answer the questions of what, where, when, who, and why the tasks are essential for engineer officers.

Specified or Implied. Is the SKB explicitly stated in current or future doctrine, regulation, or similar military authoritative sources, or is it implicitly stated as a task required to accomplish another tactical task? This criterion answers "What task must be done?" For example, a specified skill in a STP manual is "Navigate using a compass and a map."

Exemplified. Was the SKB demonstrated in previous military operations or training as being valuable, or did its absence play a significant role in events? This criterion answers, "Where has this task been applied in the past?" For example, the skill, "Use PC-based terrain analysis tools, such as TerraBase II, to create TDAs" is frequently mentioned in CALL bulletins.

Timelessness. Has the SKB proven itself useful, if not required, over the decades and centuries of military operations, or is it a proposed SKB that applies indefinitely into to future military actions? This criterion answers, "When does this task apply?" For example, the ability to "Navigate using GPS equipment" is a recently required skill, but the ability to "Determine direction without a compass" is a timeless behavior. Colonel Robert Kirby, Deputy Director of the Topographic Engineer Center, described this enduring quality as "The tools of our trade may change, but our fundamentals and purpose have not" (Piek 1998, 33).

Universality. Does the SKB apply to a majority of engineer officers at a particular rank, such as a lieutenant conducting a reconnaissance mission, or is it a niche job that most engineers would not encounter, such as a lieutenant colonel working as a deputy district commander in the Corps of Engineers (USACE)? This criterion answers, "Who must perform this task?" Lieutenant Colonel Earl Hooper, former director of the Terrain Visualization Center (TVC) in TPIO-TD, described this quality of geospatial engineering as: "It is not the mission of only a few centrally located experts (assistant Corps engineers and terrain warrant officers) within the Regiment. Engineer officers at theater, corps, division, brigade, and battalion levels must be the terrain experts . . ." (Hooper, Morken, and Murphy 2001, 15).

Tactical Fit. Does the SKB cause geospatial engineering to directly facilitate tactical operations? This criterion answers, "Why is the task important?" It takes into account the views of the engineer branch's customers--the maneuver commanders, tactical staffs, and other BOS units. For example, one of the goals of geospatial engineering is to assist the commander in visualizing the terrain. Multiple officer tasks support this goal, such as the skill "Conduct a reconnaissance" and "Conduct terrain analysis."

After evaluating each of the potential SKBs with these five criteria, an overall objective assessment will be based on the number of "yes" scores for the SKB. To be considered an essential task to aid engineer officers as terrain experts, the SKB must have at least three "yes" scores in any of the five criteria. Three of five positive scores indicate that it is a proven task in doctrine and/or practice and is a fundamental task in time and/or breadth of application. Tasks that score "yes" in only one or two of the criteria might be helpful SKBs, but are not required for the engineer officer. Justification for the "yes" scores can be found in the respective task description in Chapter 2 and in its supportive connection with other tasks. To facilitate the evaluation by the five criteria, the SKBs have been listed in Table 1 through Table 9 by the geospatial engineering functions or applications.

Several abbreviations are used in the tables in order to display as much of the task information as possible. "RANK" stands for the engineer officer ranks responsible for the task: second lieutenant (2LT), first lieutenant (1LT), captain (CPT), major (MAJ), lieutenant colonel (LTC), and colonel (COL). A plus sign after the initial rank indicates that the task applies to the given rank and all those ranks above it.

Table 1. Terrain Analysis Tasks

Potential Skills, Knowledge, and Behavior	SKB	RANK	Specified Or Implied?	Exemplified?	Timeless?	For All Engineer Officers?	Tactical Fit?	Essential SKB?
Provide input to IPB.	B	CPT+	Yes	No	Yes	Yes	Yes	YES
Conduct terrain analysis (using OCOKA).	S	2LT+	Yes	Yes	Yes	Yes	Yes	YES
Identify terrain features on a map.	S	2LT+	Yes	No	Yes	Yes	Yes	YES
Prepare a MCOO.	S	2LT+	Yes	No	No	Yes	Yes	YES
Identify weather impacts on the terrain.	K	2LT+	Yes	No	Yes	Yes	Yes	YES
Validate seasonal terrain feature data.	K	2LT+	No	No	Yes	No	Yes	NO
Understand military geography for different regions.	K	2LT+	No	Yes	Yes	Yes	Yes	YES
Use terrain to deceive the enemy.	S	1LT+	No	Yes	Yes	No	Yes	YES
Conduct terrain analysis for urban environments.	S	1LT+	Yes	No	Yes	Yes	Yes	YES
Conduct terrain analysis for peacekeeping operations.	S	CPT+	No	Yes	Yes	No	Yes	YES
Prepare engineer estimates to include geospatial engineering capabilities.	S	1LT+	Yes	No	Yes	Yes	Yes	YES
Understand enemy weapon's capabilities.	K	2LT+	No	Yes	Yes	Yes	Yes	YES
Understand how terrain affects unit camouflage operations.	K	2LT+	Yes	No	Yes	Yes	Yes	YES
Visualize the terrain.	S	1LT+	No	No	Yes	Yes	Yes	YES
Identify the critical terrain elements for a unit defensive position.	B	2LT+	Yes	Yes	Yes	Yes	Yes	YES
Evaluate terrain 360 degrees from desired locations.	B	2LT+	No	No	Yes	Yes	Yes	YES
Supervise site selection and layout.	B	2LT+	Yes	No	Yes	Yes	No	YES
Provide terrain analysis for DOCC activities.	K	CPT+	No	Yes	No	No	Yes	NO

Table 2. Data Collection Tasks

Potential Skills, Knowledge, and Behavior	SKB	RANK	Specified Or Implied?	Exemplified?	Timeless?	For All Engineer Officers?	Tactical Fit?	Essential SKB?
Evaluate the availability of standard and nonstandard map products.	B	CPT+	Yes	No	No	Yes	Yes	YES
Request standard NIMA products through logistics channels or DLA.	S	2LT+	Yes	No	Yes	Yes	No	YES
Obtain standard and non-standard terrain-products through controlled sources, such as the SIPRNET.	S	CPT+	No	No	No	No	Yes	NO
Understand the geospatial capabilities available in Joint Operations.	K	MAJ+	Yes	No	No	Yes	Yes	YES
Understand how space-based systems can enhance warfighting capabilities in reconnaissance, position and navigation, and weather.	K	MAJ+	Yes	No	No	Yes	Yes	YES
Understand the FD construct from NIMA.	K	2LT+	Yes	No	No	Yes	Yes	YES
Understand map datums and scales.	K	2LT+	No	Yes	Yes	Yes	Yes	YES
Understand "Reachback" capabilities for GI&S support in the SBCT and UA.	K	CPT+	Yes	No	No	No	Yes	NO
Conduct a reconnaissance.	S	2LT+	Yes	No	Yes	Yes	Yes	YES
Submit both verbal and written patrol reports as required by STANAG 2003.	B	2LT+	Yes	No	Yes	Yes	Yes	YES
Direct engineer reconnaissance missions.	B	2LT+	Yes	No	Yes	Yes	Yes	YES
Perform military sketching.	S	2LT+	No	Yes	No	No	Yes	NO
Record terrain information daily.	B	2LT+	No	Yes	No	Yes	Yes	YES
Coordinate with the S2/G2 and S3/G3 for collecting terrain information.	B	CPT+	Yes	Yes	No	No	Yes	YES
Establish IR for essential elements of terrain or engineer information.	B	1LT+	Yes	Yes	No	Yes	Yes	YES
Track templated and known obstacles (friendly/enemy).	S	1LT+	Yes	No	Yes	No	Yes	YES

Table 3. Data Generation Tasks

Potential Skills, Knowledge, and Behavior	S K B	R A N K	Specified Or Implied?	Exemplified?	Timeless?	For All Engineer Officers?	Tactical Fit?	Essential SKB?
Coordinate with the S2/G2 to define, prioritize, and request topographic products.	B	MAJ+	Yes	No	Yes	No	Yes	YES
Understand the organization and capabilities of the DS corps topographic engineer company.	K	CPT+	Yes	No	Yes	Yes	No	YES
Prioritize and task the production of the DS corps topographic engineer company.	B	LTC+	Yes	No	No	No	No	NO
Transmit essential terrain information to terrain teams for product update.	B	CPT+	Yes	No	No	Yes	Yes	YES
Provide the status of infrastructure for contingency operations.	S	CPT+	Yes	No	No	Yes	Yes	YES
Execute target-folder battle drills.	S	CPT+	Yes	No	No	Yes	Yes	YES

Table 4. Database Management Tasks

Potential Skills, Knowledge, and Behavior	S K B	R A N K	Specified Or Implied?	Exemplified?	Timeless?	For All Engineer Officers?	Tactical Fit?	Essential SKB?
Establish data and database management practices.	B	CPT+	No	No	Yes	No	No	NO
Disseminate terrain analysis and other geospatial products.	B	CPT+	Yes	No	Yes	No	Yes	YES
Resolve differences between reports and products to render a single COP.	B	CPT+	No	Yes	No	Yes	Yes	YES
Integrate nonstandard and non-US GI&S products into tactical databases.	K	CPT+	No	Yes	Yes	No	Yes	YES
Maintain and update the map UBL for a company or battalion.	S	2LT+	No	Yes	Yes	No	No	NO

Table 5. Data Manipulation and Exploitation Tasks

Potential Skills, Knowledge, and Behavior	S K B	RANK	Specified Or Implied?	Exemplified?	Timeless?	For All Engineer Officers?	Tactical Fit?	Essential SKB?
Use PC-based terrain analysis tools, such as TerraBase II, to create TDAs.	S	2LT+	No	Yes	Yes	Yes	Yes	YES
Understand the BTRA capabilities embedded in ABCS platforms.	K	CPT+	Yes	No	Yes	Yes	Yes	YES
Understand the capabilities of JMTK as a component of ABCS.	K	CPT+	No	No	Yes	Yes	Yes	YES
Perform grid coordinate conversions.	S	2LT+	No	Yes	Yes	Yes	No	YES
Use a digital SA overlay to conduct a map reconnaissance.	S	CPT+	Yes	No	No	Yes	Yes	YES
Use a digital SA overlay to conduct a map orientation.	S	CPT+	Yes	No	No	Yes	Yes	YES
Understand the capabilities of an engineer terrain team and the DTSS.	K	2LT+	Yes	Yes	Yes	Yes	Yes	YES
Understand the different types of digital GI&S data and their military uses.	K	2LT+	No	Yes	Yes	Yes	No	YES
Understand the digital size of standard NIMA and DTSS GI&S products for ABCS use.	K	CPT+	No	Yes	Yes	Yes	No	YES

Table 6. Cartographic Production and Reproduction Tasks

Potential Skills, Knowledge, and Behavior	S K B	RANK	Specified Or Implied?	Exemplified?	Timeless?	For All Engineer Officers?	Tactical Fit?	Essential SKB?
Manage printing and survey assistance for the rapid replication of products.	B	LTC+	Yes	No	Yes	No	No	NO
Prioritize terrain product production.	B	CPT+	No	Yes	No	No	Yes	NO
Provide bridge classification maps.	S	CPT+	Yes	Yes	Yes	Yes	Yes	YES

Table 7. Geodetic Survey Tasks

Potential Skills, Knowledge, and Behavior	S K B	RANK	Specified Or Implied?	Exemplified?	Timeless?	For All Engineer Officers?	Tactical Fit?	Essential SKB?
Understand the role of the survey platoon in the DS corps topographic engineer company.	K	LTC+	Yes	No	Yes	No	No	NO
Understand the levels of accuracy provided by geodetic survey and GPS.	K	CPT+	No	No	Yes	Yes	Yes	YES

Table 8. Terrain Advice Tasks

Potential Skills, Knowledge, and Behavior	SKB	RANK	Specified Or Implied?	Exemplified?	Timeless?	For All Engineer Officers?	Tactical Fit?	Essential SKB?
Understand and coordinate the seven geospatial engineering functions for tactical echelons.	B	2LT+	Yes	No	Yes	Yes	Yes	YES
Understand the roles of the engineer officer, terrain analysis warrant officer and topographic analyst.	K	2LT+	Yes	No	No	Yes	Yes	YES
Prepare the topographic annex or appendix to tactical plans and orders.	B	MAJ+	Yes	No	Yes	No	Yes	YES
Integrate a terrain team in brigade operations.	K	CPT+	Yes	Yes	Yes	No	Yes	YES
Identify critical logistics requirements for organic geospatial engineering support.	K	CPT+	No	Yes	Yes	No	No	NO
Serve as the corps topographic officer.	B	LTC+	Yes	No	Yes	No	No	NO
Identify threats/risks to geospatial operations and functions.	B	CPT+	No	Yes	Yes	No	No	NO
Advise the commander on the use of terrain for combat operations.	B	2LT+	Yes	Yes	Yes	Yes	Yes	YES
Assist the maneuver commander with terrain visualization.	B	CPT+	Yes	No	No	Yes	Yes	YES
Brief terrain effects.	B	2LT+	Yes	No	Yes	Yes	Yes	YES
Understand the importance of military topography and map production to senior leaders.	K	CPT+	No	No	Yes	Yes	Yes	YES
Translate tactical requirements into GI&S products and analysis.	B	1LT+	No	Yes	No	Yes	Yes	YES
Translate GI&S into tactical terms for warfighters.	B	1LT+	No	Yes	Yes	Yes	Yes	YES
Understand geospatial engineering's role in assured mobility.	K	2LT+	No	No	Yes	Yes	Yes	YES
Prepare a risk assessment when the lack of geospatial information creates uncertainty.	B	CPT+	No	Yes	No	Yes	Yes	YES
Understand geospatial information requirements of each BOS that support the tactical plan.	K	MAJ+	Yes	Yes	No	No	Yes	YES
Provide geospatial engineering advice to S4/G4 for MSR & logistics operations.	B	CPT+	Yes	Yes	No	No	Yes	YES
Understand what topographic support is available for MOOTW.	K	CPT+	Yes	No	Yes	Yes	Yes	YES

Table 9. Navigation Tasks

Potential Skills, Knowledge, and Behavior	SKB	RANK	Specified Or Implied?	Exemplified?	Timeless?	For All Engineer Officers?	Tactical Fit?	Essential SKB?
Perform map reading.	S	2LT+	No	Yes	Yes	Yes	Yes	YES
Determine the grid coordinates of a point on a military map.	S	2LT+	Yes	No	Yes	Yes	Yes	YES
Measure distance on a map.	S	2LT+	Yes	No	Yes	Yes	Yes	YES
Orient a map to the ground by map terrain association.	S	2LT+	Yes	Yes	Yes	Yes	Yes	YES
Identify topographic symbols on a military map.	S	2LT+	Yes	Yes	Yes	Yes	Yes	YES
Determine a location on the ground by terrain association.	S	2LT+	Yes	No	Yes	Yes	Yes	YES
Determine a magnetic azimuth using a lensatic compass.	S	2LT+	Yes	No	No	Yes	Yes	YES
Navigate using a map and a compass.	S	2LT+	Yes	Yes	Yes	Yes	Yes	YES
Determine direction without a compass or GPS.	S	2LT+	Yes	No	Yes	Yes	No	YES
Navigate using GPS equipment.	S	2LT+	No	Yes	No	Yes	Yes	YES

Summary

This chapter evaluated the eighty-seven potential engineer officer tasks (thirty-one skills, twenty-seven knowledge tasks, and twenty-nine behaviors) described in Chapter 2 against five criteria to determine whether or not they were essential tasks for engineer officers. From the decision matrix, seventy-three tasks (twenty-eight skills, twenty-three knowledge tasks, and twenty-two behaviors) stood out as essential SKBs. These tasks cover the seven-geospatial engineering functions and the two applications of geospatial engineering--terrain advice and navigation. Figure 4 provides a snapshot of the research progress.

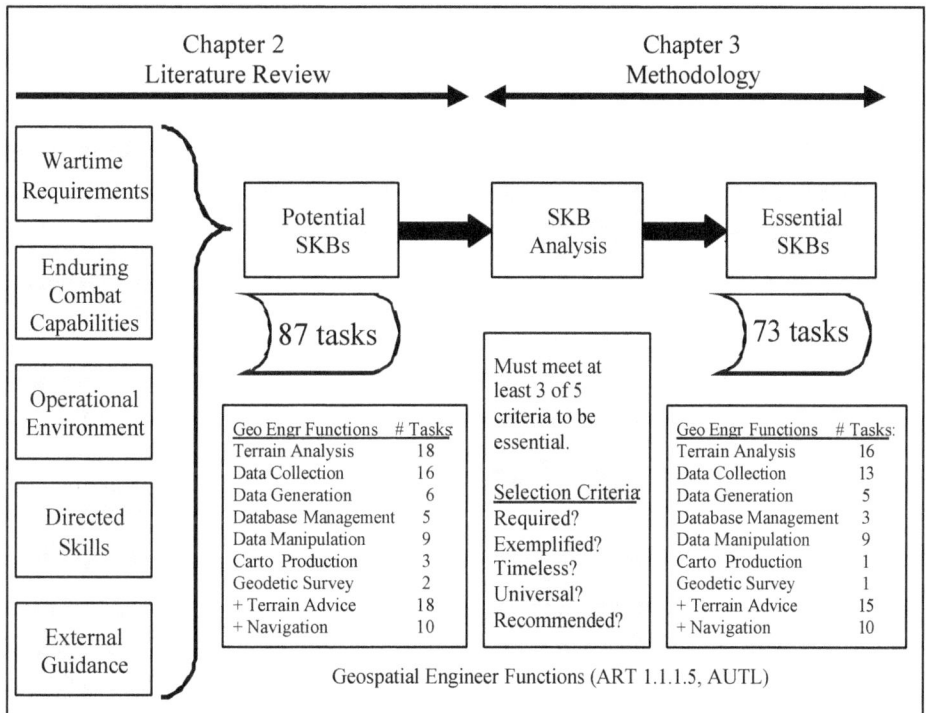

Figure 4. Essential SKB Selection Progress

CHAPTER 4

ANALYSIS

Now that the essential SKBs have been identified, this chapter reviews the findings, analyzes whether the task list is adequate, and compares the list against current institutional OES training.

Findings

This first section looks at how the essential and non-essential tasks breakdown by scoring against the five criteria, by officer ranks, and by comparison to the proposed definition for terrain expert.

When looking at the total "yes" scores for each criterion against the entire list of potential SKBs, the criteria rated highest to lowest in "Tactical Fit" (85 percent scored "yes"), "For All Engineer Officers" (70 percent), "Timelessness" (66 percent), "Specified or Implied" (66 percent), and "Exemplified" (39 percent). This gradation of the criteria scoring reflects the general scope of the study where the literary sources were tactically oriented (thus the high scoring in "Tactical Fit"), and the actual practice of terrain expertise was not as apparent (thus the low scoring in "Exemplified"). The most significant criterion between essential and non-essential tasks was "For All Engineer Officers." Essential tasks scored "yes" 84 percent of the time, while none of the fourteen non-essential tasks were applicable to the majority of officers at any given rank. These non-essential tasks belong to higher ranks (eleven of fourteen apply to captains and above), niche jobs, such as "Serve as the corps topographic officer," and to engineer terrain warrant officers and terrain analysts, such as "Validate seasonal terrain feature data."

Another perspective on the data is how it breaks out by officer ranks. As identified during the literature review, most tasks align with particular officer ranks or positions. This study assumes that once a task is learned, it is retained for possible use or at least familiarization. In many cases, a senior leader will direct and incorporate the performance of these tasks by subordinates. Figure 5 shows the cumulative growth of the SKB list over a twenty plus year career.

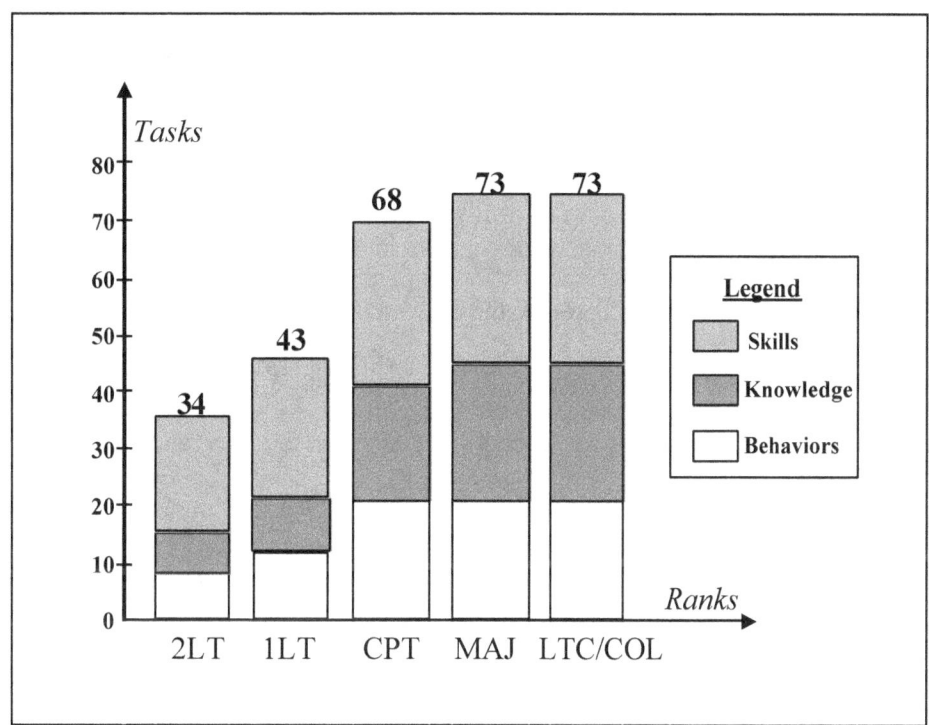

Figure 5. Essential SKB Distribution by Officer Rank

Second lieutenants have the sharpest learning curve with thirty-four essential tasks required prior to assignment to their first unit. Promotion to first lieutenant brings an additional nine SKBs. Eighteen to twenty-four months later, a captain must begin development and application of twenty-five more SKBs for a total of sixty-eight. These significant jumps in proficiency put a premium on the institutional portion of OES to educate and prepare young officers for their tactical assignments. Can the current OES system accomplish this? This question will be looked at in the third section of this chapter. Field-grade officers perform just five additional SKBs while remaining familiar with the other sixty-eight geospatial engineering tasks.

Before analyzing whether the list of seventy-three essential tasks is the right amount for an engineer officer, do the SKBs support the proposed definition for the terrain expert? From Chapter 2, a terrain expert is one who demonstrates skills, knowledge, and behaviors in rendering geospatial engineering to the tactical situation in order to take advantage of the battlespace environment. The expert understands the limits and capabilities of GI&S and can integrate them into the appropriate tactical language and processes. As shown in Chapter 3, the essential SKBs cover all seven geospatial engineering functions for officers. The SKBs are also highly supportive of the tactical situation, where 90 percent of the seventy-three essential tasks scored "yes" under the criterion "Tactical Fit." Next, at least seventeen essential tasks describe what GI&S limitations and capabilities engineer officers should be aware of to support the tactical processes. Finally, just as the role of terrain expert appears in doctrine, research found that two of every three essential SKBs (fifty of seventy-three) are stated explicitly or implicitly in doctrine. The list of essential SKBs thoroughly supports the definition for terrain expert.

Adequate?

Does this solid support for the definition come from the list having too many tasks? Could an officer be considered a true terrain expert with less tasks to accomplish? Or, is the list incomplete or too short? Are there other SKBs that were overlooked? This second section will look at the list's adequacy.

There are at least three arguments for this task list being too long--lack of task hierarchy, inclusion of nontraditional geospatial engineering tasks, and lack of prioritization. The research method did not facilitate the recognition or establishment of a clear hierarchy of supporting and supported tasks causing all tasks to be listed as equals of one another. For example, "Identify terrain features on a map" supports "Conduct terrain analysis" that supports "Provide input to IPB" that supports "Advise the commander on the use of terrain for combat operations." All four tasks were listed independently, instead of falling under the last behavior. The advantage of listing every unique task is that a myriad of hierarchies can be established using common tasks, such as " Identify terrain features on a map." This technique also allows more complex tasks, such as "Advise the commander on the use of terrain for combat operations," to be defined in greater detail, than consolidating the SKBs into a handful of key tasks. The disadvantage is managing a list of seventy-three tasks without a clear set of relationships.

Going through the list, there are at least twenty-five SKBs not traditionally thought of as geospatial engineering tasks, such as performing reconnaissance, navigation, and risk assessments. These tasks made the list because they significantly contribute to GI&S, or they heavily rely on the application of GI&S. The former category of tasks includes gathering critical terrain information through R&S planning with the S2, directing and reporting reconnaissance, and understanding the effects of enemy capabilities that affect terrain. The latter category of tasks includes embedded terrain analysis capabilities in C4ISR systems and navigation on the battlefield. These SKBs demonstrate the integral relationship of the three engineer battlefield functions. In general engineering, "Conduct terrain analysis" and the use of GI&S tools impact the performance of "Supervise site selection and layout" and "Provide the status of infrastructure for contingency operations." In combat engineering, the tasks to track obstacles, prepare engineer estimates, and evaluate enemy capabilities require understanding of how terrain affects tactics and what GI&S products best describe these effects. Inclusion of these twenty-five non-traditional SKBs underscores geospatial engineering's fundamental role in overall military engineering.

Lack of task prioritization gives the impression that all seventy-three essential tasks are equally important, creating a sense that there are too many tasks to train and master. In Figure 5, an engineer second lieutenant should be proficient in all thirty-four essential tasks during his first unit assignment. In what priority or sequence should these thirty-four SKBs be developed? The essential task development process, like the METL development process, does not prioritize tasks by design; it only identifies the critical tasks to accomplish the mission. Yet from both lists, priorities can be developed. After developing the METL, the commander evaluates the unit's readiness in each critical task and trains the unit according to its strengths and weaknesses. Officer readiness requires a similar evaluation of terrain expertise in order to develop an institutional and self-development training plan to address the strong and weak SKBs. This thesis provides the essential task list to conduct evaluations as needed at USAES, in units, and by officers.

Thus, task prioritization is not as important as task identification until a relevant assessment can be made base on COE, branch, unit, and OES factors.

One reason that it is difficult to organize the tasks by hierarchy or priority is that there is not an identified flow or process for leaders to accomplish geospatial engineering. Combat engineering has the engineer estimate, and general engineering frequently employs planning techniques, such as the critical path method and Gantt charts. Geospatial engineering is defined by the seven-geospatial engineering functions from FM 7-15, but these functions are not connected in a sequential process in doctrine. They can be arranged in the order that they might occur within MDMP as shown in Table 10, where they primarily support mission analysis. The challenge to engineer leaders is continuing the application of geospatial engineering throughout the rest of MDMP and during the tactical operations that follow.

By aligning the list of SKBs and the geospatial engineering functions to support the MDMP, a process does emerge as shown in Table 10. This alignment results in a process of five overlapping steps: assess, acquire, analyze, advise, and apply. The last step of "apply" creates the need for new assessments, and the process repeats in a loop. In fact, this process continues to loop from the first warning order of deployment through redeployment. In the table, the column for geospatial engineering functions includes the two thesis additions of terrain advice and navigation. Definitions for the five process steps follow after Table 10.

Table 10. Comparison of Processes and Functions.

MDMP (FM 101-5)	Engineer Estimate (FM 5-100)	Geospatial Engineering Functions (ART 1.1.1.5)	Geospatial Engineering Process
Receipt of Mission	Receive the Mission	Data Collection	Assess
Mission Analysis	Conduct IBP/EBA Analyze the Engineer Mission	Data Generation DB Management Data Manipulation Terrain Analysis	Acquire Analyze
Course of Action (COA) Development	Develop a Scheme of Engineer Operations (SOEO)		Advise
COA Analysis	Wargame and Refine the Engineer Plan	Terrain Advice	
COA Comparison	Recommend a COA		
COA Approval	Finalize the Engineer Plan	Cartographic Reproduction	Apply
Orders Production	Issue Orders	Geodetic Survey Navigation	

1. <u>Assess</u>. Engineer officers translate tactical missions into geospatial engineering requirements. They understand what standard products, non-standard products and topographic capabilities are available to their respective echelon. This step continues through COA comparison to identify GI&S requirements for the proposed missions of the combined arms team.

2. <u>Acquire</u>. Engineer officers know the basics of GI&S products and how to obtain them, whether through the Defense Logistics Agency (DLA), the supporting terrain team, reconnaissance, or R&S planning. As terrain experts, they assist the tactical commander by managing and continuously improving the digital (MCS and FBCB2) and analog (acetate overlay) COP. This step should occur throughout the entire MDMP, with focus on what products are needed for analysis, advice, and application.

3. <u>Analyze</u>. Engineer officers analyze the open or urban terrain and associated GI&S products for the appropriate ODSS mission. They know what products to request or generate to present this analysis. This step continues through COA comparison in MDMP, as different plans are considered.

4. <u>Advise</u>. Engineer officers translate pertinent GI&S products and analysis into tactical impacts for the commander, staff, and BOS elements. They assist the commander in terrain visualization through verbal, graphical and digital C4ISR means. They advise the commander on what friendly and enemy capabilities transform the terrain. This step continues through COA approval in MDMP.

5. <u>Apply</u>. Engineer officers integrate GI&S and analysis with available combat and general engineering assets to shape the battlespace for the tactical operation. They also use GI&S to navigate on the battlefield and conduct unit operations. This step can begin at any point during MDMP, but generally comes after a COA is approved.

Using this five-step geospatial engineering process, the essential SKBs fit into a framework that the officer can use to directly support tactical operations and add new SKBs to become a better terrain expert. In Appendix D, the seventy-three SKBs are distributed in five tables under the step they first or best apply.

On the other hand, the list of seventy-three essential SKBs might be considered inadequate because it is too short. Thesis research, covering over forty-eight primary and secondary sources, focused on US Army engineer-related material. Thus, there may be valuable SKBs from foreign armies and sister services that employ engineers as topographers that were not investigated. Also, the research methodology and selection criteria limited the development of new tasks for the Objective Force. Potential SKBs, such as "Obtain standard and non-standard terrain-product through controlled sources," often lacked the literary support to pass three of the five criteria. Even among essential SKBs, the twenty-four tasks that received "No" under the "Timeless" criterion usually did so because they involve new technologies and digital data not yet commonly used, such as "Use a digital SA overlay to conduct a map reconnaissance." So, the methodology was not as helpful in establishing new SKBs for the Objective Force. To ensure the SKB list is complete, further research should be conducted of allied and joint capabilities in terrain expertise. The omissions of potential foreign and future SKBs do not seem to render the proposed list inadequate to accomplish the terrain expert role.

The list of seventy-three essential SKBs for an officer is not as extensive as the critical task lists for a 215D warrant officer and an 81T soldier (Appendix C). Table 11 summarizes and compares the critical tasks of officers, terrain warrant officers and enlisted topographic analysts. The chart also demonstrates how officers serve as the bridge between the technical side of geospatial engineering, performed by the 215D warrant officers and 81T soldiers, and the tactical component of combined arms commanders and staff.

Table 11. Essential Tasks for 21B Officers, 215D Warrant Officers, and 81T Soldiers

Geospatial Engineering Function or Operation	21B Engineer Officer	215D Terrain Warrant Officer	81T Terrain Analyst
ASSESS	12 tasks: understand and use the digital terrain capabilities in C4ISR systems. Understand GPS use.	26 tasks: plan and manage GI&S production, support, and procedures.	38 tasks: DTSS and FBCB2 operations; 81T40 performs topographic operations management.
ACQUIRE: DATA COLLECTION	15 tasks: knowledge of GI&S products and their sources; coordinate with S2/S3 for ISR collection.	8 tasks: generate digital GI&S and exploit national sources.	12 tasks: all GI&S specific.
ACQUIRE: DATABASE MANAGEMENT	3 tasks: oversee product dissemination and maintenance of a single COP.	21 tasks: collect, catalog, evaluate, order, update, warehouse GI&S data; manage GI&S file servers.	20 tasks: GI&S data, server, and DTSS focus
ANALYSIS	23 tasks: terrain analysis for MDMP.	13 tasks: identify GI&S requirements of the MDMP and BOSs.	36 tasks: create MCOOs and other standard topo products, quality control
ADVISE	8 tasks: translate GI&S into tactical advice for the commander, staff, and BOS elements; brief.	5 tasks: plan topographic operations at operational, strategic, and joint levels; brief.	12 tasks for 81T30: prepare and brief terrain effects and visualization; manage production.
APPLY	10 tasks: navigate using maps, compass, GPS, and terrain association.		1 task: perform basic map reading.
SYSTEM ADMINISTRATION		14 tasks: terrain team operations; maintain and integrate topo hardware	6 tasks for 81T40: connect and operate DTSS on tactical LAN.
TOTAL TASKS	73 tasks	87 tasks	125 tasks

Another reason the essential task list may be inadequate is that the chosen methodology for tasks did not facilitate the incorporation of important cognitive abilities. Some of these abilities are critical in accomplishing the larger SKBs. Clausewitz cites the importance of having a good memory and good vision. He also stresses that leaders should possess a "sense of locality"--the ability to quickly and accurately grasp the topography of any area. It is the act of imagination where the terrain is "imprinted like a picture, like a map, upon the brain, without fading or blurring in detail" (Clausewitz, 1976, 109-110). Patrick O' Sullivan in *Terrain and Tactics* describes this "sense of locality" as one's spatial ability. "It seems obvious that a soldier needs to be skillful spatially in order to survive" (1991, 165). Other abilities, such as the depth perception, reading, public speaking, and computer literacy, can also significantly affect the performance of the SKBs. But when, how, and to what standard one should be able to see, memorize, imagine, and communicate has yet to be addressed for the engineer officer in the SKB list. Thus, further research is recommended into these cognitive abilities that enhance the development and application of terrain expertise.

Overall, this list appears adequate to enable engineer officers as terrain experts. The seventy-three essential tasks describe the supporting tasks that facilitate the broader geospatial engineering functions and the terrain expert definition. The many tasks fit into a simplified process to aid their career long development and continual tactical application. The list also permits task prioritization to meet relevant training and COE requirements and provides a foundation to add future SKBs too.

Existing OES Training

> To act successfully in the face of particularities of geography requires either luck or trained powers of observation and the ability to construct a mental image of the landscape. (O' Sullivan 1991, 25)

The Army and engineer branch conduct institution-level training for geospatial engineering at various career courses. All Army officers begin receiving military training as officer cadets or candidates at USMA and in the Reserve Officer Training Corps (ROTC). Once commissioned, engineer second lieutenants attend the Basic Officer Leadership Course (BOLC) at USAES. In BOLC, engineer officers receive fifteen hours of specific geospatial engineer training. Once eligible for promotion to captain, officers attend ECCC at USAES where they receive at least nineteen hours of geospatial engineer instruction. The USAES also provides distance-learning tools for all engineers through the "Terrain Visualization Compact Disk" (TV-CD). Engineer officers may also attend geospatial engineering training at the National Geospatial Intelligence School (NGS) at Fort Belvoir, Virginia. NGS offers the two-week Topographic Officers Management Course (TOMC) and one-week Geospatial Digital Data User's Course (GEODDUC). Approximately one hundred engineer officers attended these two courses from January 2001 to January 2003. NGS also provides mobile training team (MTT) classes on geospatial engineering subjects per unit request.

The full list of essential SKBs is compared against this slate of institutional training in five tables located in Appendix D. The tables are set up according to the five steps of the geospatial engineering process: assess (twelve tasks), acquire (eighteen), analyze (twenty-three), advise (eight), and apply (twelve). Similar to the nine tables in Chapter 3, tasks are listed on the left, followed by the task type (SKB), the rank at which initiated, and the sources of institutional training.

Observations on Institutional Training

USAES and NGS provide institutional training for forty-nine of the seventy-three essential tasks. The other twenty-four tasks are not in the courses available to most engineer officers. Both groups of SKBs are discussed below.

Training for Lieutenants

Precommissioning training in ROTC and USMA covers twelve and fourteen tasks respectively required by second lieutenants. All but one of these tasks ("Navigate using GPS equipment") are further taught in BOLC. Once commissioned, engineer officers receive training in twenty-nine of the thirty-four of the SKBs (85 percent) for second lieutenants and thirty-two of the forty-one tasks (78 percent) for first lieutenants at BOLC. USAES provides the TV-CD to officers at BOLC, ECCC, and field units as a tool for operational and self-development training. As the tables in Appendix D show, the TV-CD covers seventeen SKBs, eleven of which apply specifically to lieutenants. While these eleven tasks are taught in BOLC also, they reinforce SKBs commonly used in the field, such as "Use PC-based terrain analysis tools, such as TerraBase II, to create TDAs." NGS also provides several multi-media packages, such as "Geospatial Information and Services for the Warrior," for unit and soldier training. Overall, USAES appears to give engineer lieutenants a good foundation to be terrain experts for their tactical echelon assignments. But 75 percent of engineer captains (sixty-seven of eighty-eight) interviewed at ECCC felt they needed more training to succeed as the terrain experts. As one student commented, "Units want trained second lieutenants with terrain expertise, but most of my training came from the unit." While the current BOLC training covers the majority of essential SKBs, USAES should continue to investigate the operational requirements for lieutenants to refine the geospatial engineering training.

Training for Captains

Engineer captains must master twenty-seven additional SKBs beyond that of first lieutenants for a total of sixty-eight tasks. The ECCC covers only seven of these twenty-seven SKBs, but does reinforce twenty-one SKBs previously taught in BOLC. This redundant training helps refresh and improve proficiency in key tasks, such as "Conduct terrain analysis" and "Use PC-based terrain analysis tools, such as TerraBase II, to create TDAs," since captains must still be proficient in these skills. In fact, 50 percent of the young captains (forty-four of eighty-eight) interviewed felt they needed more training in terrain analysis and 58 percent (fifty-one of eighty-eight) felt they needed more training on TerraBase II and other automated tools. Due to the limited training time and resources available in ECCC, captains are not familiar with most of the new SKBs they will need. Many of these untaught tasks involve skills using new C4ISR capabilities and behaviors on how to interact with terrain teams and key staff elements for GI&S. The ECCC appears to be a weak link in keeping engineer officers prepared for the terrain expert role. The USAES is working on updating ECCC geospatial engineering classes in the coming year that should address some of the training shortfalls. It is also working on the installation and incorporation of MCS and FBCB2 systems into institutional training over the next three years (2004--2006) that will address some of the new C4ISR tasks (Granger 2003). The ECCC is the last engineer-specific career training an engineer officer receives until his attendance as a major at CGSC, so it is a critical component of OES training for terrain expertise.

Training for Majors

There is not an engineer-specific institutional training program for majors, other than what they learn through self-development training at CGSC. Majors add five new tasks for a total of seventy-three SKBs. The ECCC and the TV-CD cover one of these tasks--"Understand the GI requirements of each BOS that support the tactical plan." The other four tasks can be trained at CGSC through self-development as long as majors are aware of the requirement for them. The TOMC and GEODDUC cover four of the five tasks, but only a handful of engineer majors receive this training at NGS. Since all majors will attend resident CGSC beginning in 2005 or 2006, USAES should look for opportunities to formally train these field-grade tasks and reinforce other earlier SKBs as part of the CGSC curriculum.

Training for Lieutenant Colonels and Colonels

Similar to majors, engineer lieutenant colonels and colonels receive no formal institutional training on geospatial engineering unless they attend the two-week long PCC at Fort Leonard Wood. PCC provides a two-hour block of instruction that covers ten geospatial engineering SKBs initially taught at company-grade levels. These tasks focus on what topographic capabilities are available at all tactical levels, how tactical requirements are translated into digital GI&S products, and what advancements are made in digital data. While there are no new SKBs for the majority of these officers, they still must retain knowledge and practice of the seventy-three tasks since they are frequently responsible for the operational training and self-development plans of units. They must be familiar with the SKBs that enable lieutenants, captains and majors be terrain experts. They must ensure unit training incorporates the appropriate geospatial engineering capabilities. They are terrain experts for both the maneuver commander and their own engineer units.

The Untrained Tasks

The twenty-four tasks not covered in precommissioning, USAES, and NGS instruction represent the delta between being an Engineer officer with some geospatial engineering knowledge and being an engineer officer terrain expert. The forty-nine trained SKBs help an engineer officer conduct engineer operations and provide basic terrain analysis and advice. The other twenty-four SKBs elevate geospatial engineering to the next level of support by enabling BOS operations, enhancing staff planning, and extracting deeper analysis. The next few paragraphs describe how these untrained SKBs affect officers as terrain experts and suggest how these tasks can be trained. These twenty-four tasks are also in Appendix D where "none" is marked in the column "Institutional Training."

The first three untrained tasks under the steps of "assess" and "acquire" involve greater interaction and understanding of the terrain team in support of the tactical echelon. They are "Integrate a terrain team in brigade operations," "Transmit essential terrain information to terrain teams for product update," and "Understand the digital size of DTSS GI&S products made for ABCS use." Regardless of who controls the terrain team, engineer officers must be aware of how to best use and exploit their capabilities. The USAES does cover the task of "Understand the capabilities of an engineer terrain team and the DTSS." Armed with this knowledge, officers conduct operational training by physically meeting their supporting terrain team and developing a habitual relationship with them.

The next six untrained tasks for "acquire" could also best be exploited using the face-to-face training method with S2/G2 staff counterparts at task force level and above.

These tasks include "Resolve the differences between reports and products to render a single COP," "Track templated and known obstacles," "Establish IR for essential elements of terrain or engineer information," "Coordinate with the S2/G2 and S3/G3 for collecting terrain information," "Coordinate with the S2/G2 to define, prioritize, and request topographic products," and "Prepare a risk assessment when the lack of geospatial information creates uncertainty." Since it is difficult to simulate S2/G2 staff actions in institutional training, operational training allows engineer officers to develop working procedures with their BOS counterparts.

The last untrained task for "acquire"--"Record terrain information daily"--and seven of the eight untrained tasks for "analyze" all enhance engineer officers' performance and output in terrain analysis. The first task should be a common behavior of all engineer officers; they learn what to record from institutional training and then practice how to record from their units. Two tasks that broaden the scope of terrain analysis include "Conduct terrain analysis for urban environments" and "Conduct terrain analysis for peacekeeping operations." Five tasks that enhance the level of detail of terrain analysis include "Evaluate terrain 360 degrees from desired locations," "Use terrain to deceive the enemy," "Understand how terrain affects unit camouflage operations," "Understand military geography for different regions," and "Provide bridge classification maps." These tasks should be trained as part of the BOLC and/or ECCC instruction on terrain analysis. They could also be taught through distance-learning packages for self-development training.

The next four untrained tasks for "analyze," "advise," and "apply" arise from GI&S in the Objective Force and new C4ISR capabilities, where institutional training is not fully developed. These tasks include "Understand geospatial engineering's role in 'assured mobility'," "Understand the BTRA capabilities embedded in ABCS platforms," "Use a digital SA overlay to conduct a map reconnaissance," and "Use a digital SA overlay to conduct a map orientation." The first two knowledge tasks dealing with assured mobility and BTRA are relatively new for engineer officers and need to be defined in doctrine. The training of the latter two tasks depend how soon USAES can integrate MCS and FBCB2 systems into BOLC and ECCC instruction. For the immediate future, officers must learn to use these systems through on-the-job training. The Combined Arms Service Staff School (CAS3) and CGSC at Fort Leavenworth expose students to these C4ISR systems as of 2002. Once the Army defines the Objective Force in detail, training on assured mobility and BTRA can be finalized.

The last three untrained tasks for "advise" and "apply" require a blend of institutional and operational training. They include "Provide geospatial engineering advice to the S4/G4 for MSR and logistics operations," "Provide the status of infrastructure for contingency operations," and "Supervise site selection and layout." Engineer officers must interface with their BOS and staff counterparts to best serve the diverse GI&S needs in tactical operations across the full-spectrum of ODSS. This includes visiting their tactical operation centers to better understand their requirements. To supply the right terrain expertise, engineer officers must also know how to access and interpret additional layers of GI&S.

Overall, of the twenty-four untrained tasks: seven should be trained by a combination of USAES instruction and field application, eight should be trained by USAES and reinforced through self-development, and nine should be trained in units.

The full list of these twenty-four SKBs with their recommended training strategy are in Table 12 in the next chapter.

Summary

This chapter reviewed the findings of the chosen methodology and results, the adequacy of the essential task list, compared this list against available institutional OES training, and made recommendations for develop the untrained tasks. The seventy-three essential tasks fully support the proposed definition for terrain expert and requirements of engineer officers at tactical levels. More importantly, the SKBs lend themselves to a geospatial engineering process presented as assess, acquire, analyze, advise, and apply. This process facilitates the incorporation of geospatial engineering throughout the MDMP and mission execution. It also gives engineer officers a framework that they can add the SKBs throughout their career to be terrain experts. Finally, the essential skills were compared with available OES institutional training. This comparison showed that ROTC, USMA, and BOLC provide a strong foundation of geospatial engineering tasks for lieutenants that should be continued. The ECCC training, while reinforcing the SKBs learned as a lieutenant, provide training for only seven of the additional twenty-five tasks for captains and one of the five tasks for majors. In total, BOLC and ECCC provide training for thirty-nine of the sixty-eight company grade SKBs (57 percent). Only NGS provided substantial training for the five field-grade officer tasks (three of five). The review of OES training also indicates that twenty-four tasks are not covered at all by USAES or NGS, and it is these SKBs that appear to make a significant difference between officers being involved with terrain or being terrain experts.

CHAPTER 5

CONCLUSIONS AND RECOMMENDATIONS

Conclusions

Terrain is a permanent and influential component of all military operations. Military theorists and leaders of yesterday and today recognize that the side who gains mental and physical dominance of the terrain can and will win. Yet, terrain is often far more complex than meets the eye or is portrayed by a map. Dominating it requires additional study and analysis in geospatial engineering, and engineers have that responsibility for the US Army. Engineer leaders are the commanders' local guides; they provide the knowledge and tools of all three engineering battlefield functions so that the commander can wield the ground as a weapon against the enemy and as a combat multiplier for the friendly forces. Therefore, just as engineer officers must be the combat engineer and general engineer experts, they must also be the commanders' terrain experts. A terrain expert is one who demonstrates SKBs in rendering geospatial engineering to the tactical situation in order to take advantage of the battlespace environment. The expert understands the limits and capabilities of GI&S and can integrate them into the appropriate tactical language and processes.

After defining the terrain expert role, the thesis described the SKBs required for all company and field-grade engineer officers. The Army's METL development process provided a sound methodology to extract essential SKBs from existing and proposed requirements for engineer officers. Eighty-seven potential tasks emerged from numerous primary and secondary sources. Using five criteria that addressed the need for who, what, where, when, and why of each task, seventy-three SKBs emerged as essential: twenty-eight skills, twenty-three knowledge tasks, and twenty-two behaviors. These SKBs accumulate over the engineer officer's career, beginning with thirty-four essential tasks as a second lieutenant and ending with seventy-three tasks as a colonel (Appendix D).

To better grasp how these seventy-three SKBs work in concert, geospatial engineering should be viewed as a process that supports the Army's MDMP and Objective Force's quality of firsts. A five-step continuous process that emerged from the selection of the SKBs is assess, acquire, analyze, advise, and apply geospatial engineering. Using these five steps as a professional framework, officers add SKBs to enhance their ability to perform this process, instead of randomly collecting a disjointed set of geospatial engineering tasks.

Training seventy-three tasks is a significant, but not overwhelming challenge for the leader development program. The burden primarily falls on institution training at USAES. As described in Chapter 4, USAES made significant strides toward this goal since the officer MOS consolidation in 1996. For example, in BOLC, second lieutenants receive training in thirty-two of the forty-one lieutenant tasks they may have to perform. This training should be continued and refined as more information and capabilities for the Interim and Objective forces emerge. The USAES can focus its efforts on improving training for captains and field grade officers in ECCC, CGSC, PCC, and distance learning media. This will enable engineer captains to master twenty-seven additional SKBs, seven of which are currently covered in ECCC. At the same time, USAES can update course hours to incorporate the eighteen missing captain tasks and the five field-grade tasks. The USAES should continue to publish distance-learning software, such as the TV-CD. Both USAES and NGS can increase cooperation to train SKBs in the institutional and operational settings by attending each other's annual training summits and through joint development of future distance-learning software.

The need for additional training is significant when looking at the twenty-four essential SKBs that are not taught by precommissioning sources, USAES or NGS. These twenty-four SKBs will facilitate engineer officers in their quest to become effective terrain experts. Many of the other forty-nine taught tasks deal with subjects (terrain analysis, map reading, navigation, ordering GI&S products, and reconnaissance) that are not unique to engineer officers. Other branch officers can and do perform them. The twenty-four untrained tasks are not so general. They often involve tasks that specifically enable engineer functions and BOS applications.

This study did not evaluate current operational and self-development training conducted by engineer units, though many engineer brigades and battalions have such programs. It is hoped that this definitive list of essential tasks facilitates unit and individual instruction as it helps USAES and NGS to better equip officers. Comments from all three interview groups strongly recommended more "boots-on-the-ground" training, where engineer officers walk the terrain with maneuver and BOS counterparts to better grasps how to analyze, interpret, and shape the terrain tactically. Simple exercises, such as tactical exercises without troops (TEWTs) and terrain walks, can improve an engineer officer's mastery of geospatial engineering tasks.

This research study concludes that there are definable and achievable SKBs to enable engineer officers as terrain experts for the tactical levels of the Army. Engineer officers can use these SKBs to improve the commanders' vision of the battlefield in the Current, Interim, and Objective forces. Trained and ready in the leader tasks of geospatial engineering, they will surpass yesterday's topographic engineers and become more complete engineer officers.

Recommendations

As the research focused on the individual leader tasks for officers in the engineer branch, the following seven recommendations are provided to the USAES commandant for consideration and potential inclusion in doctrine and training.

First, USAES should adopt a definition and set of SKBs for terrain expert to improve acceptance and application of geospatial engineering by the branch. This will help eliminate confusion among engineer and non-engineer leaders concerning this important role. The proposed terrain expert definition and essential SKBs from this research provide an example of how this role can be described to facilitate leader development and maneuver support, both now and into the Objective Force. The chosen definition and tasks should be included in the new FM 3-34 *Engineer Operations*, FM 3-34.230, *Topographic Operations*, and the applicable STPs, FMs, and JPs that reference to terrain expertise. All institutional training (resident or distance learning) on geospatial engineering at USAES and NGS should also present a common scope and duty of the engineer roles in geospatial engineering. This delineation of officer, warrant officer and soldier roles can be expressed in a simple chart, similar to Figure 3, "Geospatial Engineering Roles." This type of chart should be included in FM 3-34 and institutional training that covers geospatial engineering subject material.

Second, USAES should promote geospatial engineering as a process rather than just a subject matter to facilitate the education and incorporation of these tasks. This research study determined that the officer tasks follow a sequential logic to support Army tactical processes, such as MDMP and Objective Force quality of firsts. The five-step process is recommended for leader application of geospatial engineering: assess, acquire, analyze, advise, and apply. The USAES should incorporate this process both in doctrine and in instruction for engineer officer training.

Third, USAES should update ECCC in the following four areas so captains can be better prepared for the next level of demands for terrain expertise. One, expand terrain analysis from strictly an open terrain application of OCOKA to a more tailored analysis of complex urban environments and MOOTW missions. Two, provide tactics, techniques and procedures on how to coordinate with the S2/G2 for terrain information in R&S operations and from topographic sources. Three, provide greater interaction with terrain teams at USAES, either by using full sets of common terrain team products for course exercises or by inviting current terrain team members to speak to ECCC classes. Four, incorporate the common digital C4ISR systems, such as DTSS, MCS, and FBCB2, into institutional training at BOLC and ECCC. The previous USAES Commandant, Major General Aadland, highlighted this need:

> To maintain our relevance on the battlefield, we must develop and field a world-class engineer MCS [MSC-E]. The MCS-E is an automated decision support and management element to be embedded in MCSs supporting the maneuver commander and providing the engineer commander rapid answers other wise requiring time-consuming manual calculations. The MCS-E is tied to the DTSS and the rest of the Army Battle Command System (ABCS) and is fully integrated throughout the battlespace. (Aadland and Allen 2002, 9)

Fourth, USAES should establish the task, conditions, and standards for all essential SKBs. Special focus should go to the identified twenty-four untrained tasks. Table 12 provides the recommended training strategy for each. Operational, or unit, training is normally conducted with the aid of standing operating procedures (SOPs).

Table 12. Training Strategy for Untrained SKBs

Task	SKB	Geospatial Engineering Process	Recommended Training Strategy
Integrate a terrain team in brigade operations.	K	Assess	Familiarize at USAES and NGS Practice in unit using SOPs
Transmit essential terrain information to terrain teams for product update.	B	Acquire	Practice in unit using SOPs
Understand the digital size of DTSS GI&S products made for ABCS use.	K	Acquire	Familiarize at USAES and NGS Practice in unit using SOPs
Resolve differences between reports and products to render a single COP.	B	Acquire	Practice in unit using SOPs
Track templated and known obstacles (friendly/enemy).	S	Acquire	Practice in unit using SOPs
Establish IR for essential elements of terrain or engineer information.	K	Acquire	Practice in unit using SOPs
Coordinate with the S2/G2 and S3/G3 for collecting terrain information.	B	Acquire	Practice in unit using SOPs
Coordinate with the S2/G2 to define, prioritize, and request GI&S products.	B	Acquire	Practice in unit using SOPs
Prepare a risk assessment when the lack of geospatial information creates uncertainty.	B	Acquire	Familiarize at USAES and NGS Practice in unit using SOPs
Record terrain information daily.	B	Acquire	Familiarize at USAES and NGS Practice in unit using SOPs
Conduct terrain analysis for urban environments.	S	Analyze	Familiarize at USAES and NGS Reinforce through self-development
Conduct terrain analysis for peacekeeping operations.	S	Analyze	Familiarize at USAES and NGS Reinforce through self-development
Evaluate terrain 360 degrees from desired locations.	B	Analyze	Familiarize at USAES and NGS Reinforce through self-development
Use terrain to deceive the enemy.	S	Analyze	Familiarize at USAES and NGS Reinforce through self-development
Understand how the terrain affects unit camouflage operations.	K	Analyze	Familiarize at USAES and NGS Reinforce through self-development
Understand military geography for different regions.	K	Analyze	Familiarize at USAES and NGS Reinforce through self-development
Provide bridge classification maps.	S	Analyze	Familiarize at USAES and NGS Reinforce through self-development
Understand geospatial engineering's role in assured mobility.	K	Analyze	Familiarize at USAES and NGS
Understand the BTRA capabilities embedded in ABCS platforms.	K	Advise	Familiarize at USAES and NGS Practice in unit using SOPs
Use a digital SA overlay to conduct a map reconnaissance.	S	Apply	Familiarize at USAES and NGS Practice in unit using SOPs
Use a digital SA overlay to conduct a map orientation.	S	Apply	Familiarize at USAES and NGS Practice in unit using SOPs
Provide geospatial engineering advice to S4/G4 for MSR & logistics operations.	B	Advise	Practice in unit using SOPs
Provide the status of infrastructure for contingency operations.	S	Advise	Practice in unit using SOPs
Supervise site selection and layout	B	Advise	Practice in unit using SOPs

Fifth, the MQS manual, or future equivalent, for engineer lieutenants and captains should include the new geospatial engineering SKBs. The older manual, STP 5-21II-MQS, *Military Qualification Standards II Engineer (21) Company Grade Officer's Manual*, contains an excellent layout of task, conditions and standards for training in institutional, operational, or self-development environments.

Sixth, the EBA should include assessments of friendly and enemy geospatial engineering capabilities in addition to the combat and general engineering capabilities. This assessment could be as simple as addressing the five-steps of the proposed geospatial engineering process above for both sides.

Seventh, instruction at USAES and NGS should remind students that no GI&S product replaces the value of seeing the terrain in person. When possible, institutional training should get students out of the classroom and into the field. Captain Swift captured this concept as an essential component of officer training over one hundred years ago.

> But it must be confessed that the best map gives a very poor picture of the ground. We only accept it as a guide in the full darkness, to be supplemented by the real ground drawn in full day upon the human retina. Hence the time has arrived when we may advance another step in our career and solve our military questions on the ground itself.... (1897, 268)

Terrain walks, reconnaissance, land navigation, staff rides, field training exercises and road marches all provide invaluable opportunities for officers to better understand the realities of mixing tactics with the environment. Even a drive through urban areas can provoke serious consideration to the challenges for military operations. The USAES should consider how to reinforce geospatial engineering through these training events when available.

Recommended Areas for Further Study

Although this research tapped into dozens of different sources to uncover eighty-seven potential officer tasks, it barely scratched the surface of geospatial engineering's value and impact on the engineer branch and Army. During the study, there were four additional topics that stood out for continued study to enable engineer officers as terrain experts.

First, there appears to be unique cognitive abilities required for geospatial engineering. Softer skills, such as spatial reasoning, photographic memory, sketching, automation use, depth perception, and sensory integration seem to benefit and influence terrain expertise. If special abilities are a prerequisite for leaders, further study should be conducted to explore what these skills are and how to develop them in future engineer officers.

Second, further study should be conducted of terrain visualization to determine in what exactly commanders and engineer officers should be able to see when they visualize. The research should include evaluating what GI&S products best facilitate this form of visualization.

Third, the essential SKBs presented in this thesis need to be defined in terms of conditions and standards. This study identified what tasks are essential for the terrain expert role but did not define their objective requirements.

Fourth, the cognitive and physical effects of migrating from analog to digital GI&S products in ABCS should be identified and measured. Users must understand the limitations of various scales, datums, grid coordinates, and sources of analog and digital

GI&S products. Research should be conducted on how to help leaders of all branches improve their geospatial reasoning skills.

Enabled with these essential SKBs, engineer officers will be the commanders' local guides as terrain experts in the Current, Interim and Objective forces to dominate the battlefield. As Sun Tzu said, ". . . if you know Heaven and Earth, you may make your victory complete" (2002, 81).

APPENDIX A

COMPARISON OF ENGINEER COMPANY MTPS

Engineer Company MTP Collective Tasks	HEAVY DIV.	AR. CAV. REGT.	ABN. DIV.	LIGHT INF. DIV.	AIR ASLT. DIV.	CORPS WHEEL
Identify topographic support requirements. (05-2-1389.05-R01D)	YES	YES	no	no	no	no
Conduct a tactical reconnaissance. (05-2-0414.05-R01D)	YES	no	YES	YES	YES	YES
Conduct an engineer reconnaissance. (05-2-0407.05-R01D)	YES	YES	no	no	no	no
Conduct a technical reconnaissance. (05-2-0412.05-R01D)	YES	YES	YES	YES	YES	YES
Prepare an obstacle plan. (05-2-0001.05-R01D)	YES	no	no	YES	YES	YES
Prepare an engineer estimate. (05-2-0002.05-R01D)	YES	YES	YES	YES	YES	YES
Conduct a convoy. (07-2-1301.05-T01D)	YES	YES	YES	YES	YES	YES
Conduct engineer-intelligence collection. (05-2-0413.05-R01D)	YES	YES	no	no	no	no
Analyze battlefield information. (05-2-0415.05-R01D)	YES	no	YES	YES	YES	no
Conduct breaching operations. (05-2-0114.05-R01D)	YES	no	YES	YES	YES	YES
Execute target-folder battledrills. (05-2-1390.05-R01D)	YES	no	no	no	no	no
Fight as infantry. (05-2-1215.05-T01D)	YES	no	YES	YES	YES	YES
Establish company defensive position. (07-2-0414.05-T01D)	YES	YES	YES	YES	YES	no
Conduct report procedures. (05-2-1218.05-R01D)	YES	YES	no	YES	no	No
Camouflage vehicles and equipment. (05-2-0301.05-R01D)	YES	YES	YES	YES	YES	YES
Conduct a water-crossing and site-approach reconnaissance. (05-2-0403.05-R01D)	YES	YES	no	no	no	no
Perform EBA. (05-2-0027.05-R01D)	no	no	YES	YES	YES	YES
Fight as engineers. (05-2-1215.05-T01D)	no	YES	YES	YES	YES	YES

Engineer Company, Heavy Division--ARTEP 5-335-65-MTP
Engineer Company, Armored Cavalry Regiment--ARTEP 5-113-35-MTP
Engineer Company, Airborne Division--ARTEP 5-027-35-MTP
Engineer Company, Light Infantry Division--ARTEP 5-157-35-MTP
Engineer Company, Air Assault Division--ARTEP 5-217-35-MTP
Engineer Company, Corps Wheeled--ARTEP 5-427-35-MTP

APPENDIX B

INTERVIEW QUESTIONS AND RESULTS

This appendix contains three sets of questions used for interviews with engineer and non-engineer officers attending PCC, CGSC, and ECCC. The raw results are in the box immediately following the question, except for the open comments given for the last question of each interview. The open comments are sorted by common topics; duplicate comments were omitted.

Interview Questions for Future Battalion and Brigade Commanders

Number of interviewees: forty-three colonels and lieutenant colonels

1. What is your branch association: combat, combat support or combat service support (CSS)?

COMBAT	COMBAT SUPPORT	CSS	TOTAL
26	11	6	43

2. What types of units did you previously command or support as a primary staff officer at division level and below to include joint task forces (JTF)?

HEAVY/MECH	LIGHT	WHEELED	JTF	SPECIAL OPNS
34	29	3	11	5

3. What individual or group most often provided the terrain expertise (products, data, analysis, advice) to you and your staff?

PROVIDER	COMBAT	COMBAT SPT	CSS	TOTAL
S2/G2	18	3	4	25
Engineer officer	2	3	1	6
Terrain team	2	3	1	6
Engineer officer and terrain team	3	0	1	4
BOS staff	0	1	0	1
Not an issue	1	0	0	1

4. On a scale of one to five, with one being no support to five being exceptional support, how well did engineers provided needed terrain, or geospatial information to you and your staff?

HOW WELL?	COMBAT	COMBAT SPT	CSS	TOTAL
1. Don't provide it	7	0	2	9
2. Little terrain support	2	2	0	4
3. Some, expect more	8	3	1	12
4. Adequate expertise	8	5	2	15
5. Exceptional support	1	1	1	3

5. What are the five-geospatial products that are most important to you at division level and below to accomplish missions across the full-spectrum of operations?

PRODUCTS	COMBAT	COMBAT SPT	CSS	TOTAL
Satellite/aerial imagery	21	7	2	30
Detailed terrain analysis	19	6	1	26
Paper map products	16	4	5	25
3-D views of terrain	16	6	1	23
Digital map products	9	8	2	19
Enemy terrain usage	9	4	2	15
Airfield & port data	5	3	6	14
Weather impact analysis	9	2	3	14
LOC data	5	3	3	11
Urban-terrain data	4	3	1	8
Terrain reconnaissance	6	0	1	7
Elevation/contour data	1	5	0	6
UAV images	3	1	0	4
Vegetation data	2	0	0	2
Soil conditions	0	0	1	1

6. As the Army moves toward the Objective Force where greater mobility and information dominance will be necessary to "see first, understand first, act first and finish decisively", do you expect that . . .

   a. Less terrain expertise will be necessary due to improvements in C4ISR capabilities to display the battlefield and more information being available.
   b. About the same level of terrain expertise will be required due to the complexity of the terrain and greater resolution of terrain data needed.
   c. More terrain expertise will be necessary to collect, interpret, analyze and disseminate geospatial information in the Objective Force.
   d. Not sure.

HOW MUCH SPT?	COMBAT	COMBAT SPT	CSS	TOTAL
Less support	0	1	1	2
Same level of support	8	0	0	8
More support	17	9	5	31
Not Sure	1	1	0	2

Results: Thirty-one officers (72 percent) felt more terrain expertise is needed. Thirty-nine (91 percent) felt the same or more terrain expertise was needed in the Objective Force.

7. Do you have any other comments with regards to geospatial engineering support engineers should provide to commanders, staffs, and BOS elements for the Current, Interim, and/or Objective forces? [The raw comments from interviews are grouped as shown]

Products
- Need timely products that are timely dispersed (real-time products).
- Focus on providing the products that will help the commander make decisions early.
- Focus or tailor products for company commander use. Less details for battalion and higher.
- Continue three-dimensional products to support attack helicopter operations.
- Make products available to subscribers automatically. List product requests and available products on a web-server.
- Develop an archival database that all units can influence and access before and during combat operations.
- Digitize the geospatial products for electronic media to the soldier level.

The Engineer Officer's Role
- Take engineer lieutenants and captains out on the ground with the maneuver officers, with products so they can better visualize.
- Push an engineer officer to battalion level maneuver unit.
- Use digital, three-dimensional terrain analysis programs (Falcon View, Top Scene, etc . . .).
- There is a world of excellent digital terrain tools that are very capable. Get them, learn them, and use them.
- Teach the engineer company executive officer that he and his men are the experts in the battalion and TF tactical operations center (TOC). They need to be the experts.
- Engineers must get digital. It makes it easier to pass products down to the lower levels.
- Ensure the engineer representative is a full partner in the BOS planning and execution. Teach commanders and staffs at local levels on what you have and what support you can provide. Touch each unit.
- Every BCT needs a terrain team in their TOC.
- Train engineer officers and non-commissioned officers at company-grade level on the geospatial tools.

- Train field-grade engineer officers on how to use and/or integrate into combined arms plan.
- Educate senior engineer commanders on what's available, both organic in their units, at division/corps (terrain teams), and via "Reachback" to the Maneuver Support Center (MANSCEN) or other strategic level resources (and how to access it).
- Sponsor leader terrain walks as officer professional development (OPD) at CTCs or on home station installation terrain.
- Advertise engineer capabilities more.
- Have ADE or engineer brigade representatives spend time in various unit TOCs to see what products would enhance their operations.
- Develop "marketing or promotional" campaign plan. Show customers the type of products that you have and how these products can support decision-making.
- Engineer must understand missions of BOS specific units to understand their needs and what products are most useful. CSS units: road networks, impacts of weather or surfaces, enemy terrain uses in BCT or division rear area are critical for survival and success.
- Keep providing quality terrain expertise at all levels to include joint operations.
- Integrate engineer operations with DISCOM MMC/SPT Operations for logistics preparation of the battlefield planning.

Engineer Officer Versus the S2 Officer
- Give terrain analysis to the engineer branch - take it away from the S2s.
- In BCT and TF operations, S2s did most of the terrain products. However, at corps level, the engineers provided extensive and superb terrain data and products--much of it for real-world operations.
- I have found that I can get almost any product I wanted to support operations. However, either the cost was too high or the wait was too long for the product to be of use to me. The reason the intelligence-world and engineer-world continuously argue over responsibilities in this area is almost always a question of resources. For example, "I'm not manned to do this," or "Our budget won't support it."
- Commanders still turn to the S2/G2 for answers on how the enemy and friendly troops operate in the area of operations. While the engineer terrain teams provide terrific technical expertise, it is the S2s who have the analytical bundle. Terrain teams have got to work for the S2/G2!
- Exploit expertise of terrain teams.

Hardware (HW) and Software (SW) Issues
- Better integration of software products that provide 3-D visualization of terrain.
- HW (PCs and printers) must match the SW requirements and capabilities. Either issue the correct HW with demanding SW, or keep SW limited and packaged smaller so that it does not overwhelm HW.
- More terrain expertise is needed because FBCB2 and other digital products are on flat screens, so there is less geospatial focus by leaders.
- Continue to invest in imaging assets and automated terrain tools that give the division commander real time or near real time view of terrain.

Interview Questions for Engineer Majors at CGSC

Number of interviewees: twenty-one majors

1. On a scale of one to four, one being unfamiliar and four being very knowledgeable, how familiar are you with the digital and hardcopy terrain products available to a division or brigade?

1 - Unfamiliar	2 - Familiar, but cannot really use	3 - Know enough, but should know more	4 - Know more than enough	Average
1	9	10	1	2.52

2. On a scale of one to four, one being unfamiliar and four being very knowledgeable, how familiar are you with the available topographic companies and terrain detachments that support tactical echelons?

1 - Unfamiliar	2 - Familiar, but cannot really use	3 - Know enough, but should know more	4 - Know more than enough	Average
2	11	4	4	2.48

3. Based on the scale below, how helpful were BOLC, ECCC and other officer education in helping you understand and use geospatial engineering?

1 - Not at all	2 - Helpful, but cannot really use it	3 - Know enough, but should know more	4 - Know more than enough	Average
7	11	3	0	1.81

4. Have you ever been put in a position of leader or staff officer where you had to provide the terrain expertise for a tactical exercise or operation?

Yes--16	No--5

5. What suggestions or lessons learned for geospatial engineering support do you think other engineer officers should be taught or know about?

<u>Products</u>
- Learn what types of products/data are available.
- As an OC at the NTC, too often products would be developed but the information would not get distributed to the echelon where it was truly needed. Along the same lines, products created, without adequate analysis attached, where next to useless.
- Engineer officers need to facilitate getting that information to the planners, decision makers, and executors in as timely a manner as possible.

- Here is the problem. We go around telling everyone about all of these great terrain products that we can provide but we don't have access to them and they never seem to get in the hands of the people who actually need them on the ground. If the engineers are truly the terrain experts for the Army, then please give us the tools to supply useful products to maneuver commanders.
- Engineer need to have a good feel for the terrain analysis products that assist engineer planning phase and the IPB so that the engineer estimate makes sense. The engineer estimate without the IPB products lacks refinement.
- At our level, how to get the terrain data for any AO? What agencies can you get it from? Who/what is available to conduct technical training for our units?
- The current move toward more laptop-friendly and capable terrain products is the right start.
- We need to sell the elevation, slope, visibility, trafficability, and route data kinds of info more and get away from the gee-whiz, fly-through type things that aren't really value added except for the most mentally challenged commanders.
- Most important, where do we get the more mission specific data? How do we construct the overlays that focus commanders on specific terrain features (roads, rivers).

Automated Terrain Tools
- Engineer officers must be competent in using Terrabase II and more importantly, know how to find/use the databases and resources to put into Terrabase II for analysis.
- As an ABE, I taught my driver the basics of Terrabase II, and he was very valuable in providing me with basic products during the planning phases of operations.
- I frequently and extensively used Terrabase II as a TFE ('96--'98) but usually had my company XO produce the products based on my guidance or requests from other company commanders, TF FSO, etc . . . .
- Better familiarization and basics in using Terrabase II. Also, distribute user guides, software, websites, etc . . . .
- As an ABE at CMTC, I used Terrabase II in conjunction with the electronic intercept information was able to confirm or deny enemy positions based on inter-visibility lines and SIGINT and was expected to have a working knowledge of the Terrabase II program.
- Every maneuver commander who has resources with Terrabase II products has always wanted more. We have only had spotty success in supplying officers who can produce a consistent product across the engineer force.
- We know there are a lot of tools out there to support maneuver commanders but very few of us ever get to touch or apply them.

Applications
- As a first lieutenant (pre-ECCC) I was a battle captain with the 555th Engineer Group. I analyzed ten potential crossing sites and recommended two for the 1st Cavalry Division to cross. It was a humbling experience.
- The system is only so good. The maps do not replace boots on the ground. Many terrain variations cannot be reproduced on most maps. Even with near real-time data,

the level of detail is less than optimal. The other question is bandwidth--we can't get near real-time information without bandwidth.
- How do we use it [geospatial engineering support] in the field environment. It is one thing to learn it in a sterile classroom and another to apply it in a TOC or planning cell in an exercise.
- Be able to brief effects, not just the hard [raw] data.
- Must work hard to build credibility with maneuver commanders and staffs.
- How to paint the picture for a commander.

Capabilities
- Engineer officers need to know the capabilities of the division terrain teams and take the lead in getting them integrated into the MDMP at division and brigade.
- Division terrain teams need to be more forthcoming on providing "push packages" of the most sought-after items. The British topographic teams do a very good job at this. Many of the warrant officers I have worked around (which is the wrong word unfortunately) appear hesitant to do this, not understanding that it will cost them less in the long run in terms of time, materials and effort. These push packages should include hard and soft copy items, as well as CD-ROM terrain data that can be uploaded onto whatever platforms are carrying terrain software.
- Know how to integrate non-Army topographic capabilities with division and corps staffs (such as a NIMA support team).

Preparation
- Plan ahead. Once you hit the field you are in the react mode.
- Know what is available and how to access it.
- Suggestions from an engineer OC at NTC:
    1. Know the capabilities of the system.
    2. Use it at home station and make an SOP of products that you provide the battalion, TF or BCT (if at that level).
    3. Have redundancy in technical experts/systems (those that can produce your SOP products such as line of site and range fans.
    4. Encourage other BOS reps to use it/load on their systems (S2, SIGO, NBC, ADA and teach them to make their own products).
    5. Attach a timeline to the SOP products. Most need to be done during the IPB (EBA) but what do you need to have for wargaming and what will go out to subordinate units to help them with their planning.

Training
- How about a web base course for all engineers covering the basic tools and products?
- CTCs can create unrealistic data/product solutions since so much of it is public "xerox-centric" cookbook data.
- Pay attention at ECCC until they kill it in favor of distance learning (current instruction is very good), then when it is made, take the Geospatial Engineer DL course.

- Need to do better integration in exercises, especially at CGSC and branch CCCs-- have a terrain team or their products and a role-player to provide the accurate support.
- If USAES believes that it [all that GI&S and Terrabase II computer stuff] is the way of the future, then they need to teach it in BOLC and ECCC and encourage brigade commanders in the field to push it.
- This needs to be something that we concentrate on in our schools. Need some kind of sustainment training for units in the field. Maybe some kind of traveling team that teaches units.
- Basic individual training provided for all ranks through at Fort Leonard Wood would be very helpful.
- Need a much better grounding in the automated NIMA type training provided to the topographic folks.
- The training at TOMC focused on what was available . . . not necessarily how to best use it to support a task force.
- We get just enough training at BOLC and ECCC to be dangerous.

<u>Reputation</u>
- Be cautious about signing up for the terrain expert. As combat engineers, we provide a unique perspective and insight into terrain and how it affects operations. However, I have never bought into the concept that we are the terrain experts. I think that is more of a political statement rather than a valid one.

Interview Questions for Engineer Officers at ECCC

Number of interviewees: eighty-eight captains.

1. What type of engineer unit(s) have you served in?

Mech	IBCT	Light	CH/CSE	Topo	Basic	USACE	Other
48	4	18	17	4	4	2	1

Mech – mechanized; CH – combat heavy; CSE – combat support equipment

2. What positions have you served in prior to ECCC?

Platoon Leader	Assistant Staff	S2	XO	TFE	ABE	ADE	Other
83	39	10	54	11	10	3	6

3. Who was primarily held responsible for terrain expertise and providing terrain information in your unit and/or in the supported maneuver units?

S2	Engineer Officer	Terrain Team	BOS	N/A	Other
46	26	6		9	1

4. How many times did you provide terrain expertise to a commander, staff or other BOS, to include input for terrain analysis, terrain advise for mobility, countermobility, and survivability tasks, terrain evaluation of a planned operation or mission, or terrain products?

N/A	Never	1--2 times	3--5 times	More that 5 times
13	23	18	6	28

5. How many times have you directly worked with a division terrain team or detachment at either home station, during exercises, or on deployment?

Never	1--2 times	3--5 times	More than 5 times
55	21	7	5

6. On a scale of one to five, how well do you feel the training and training tools (such as TerraBase II) at the USAES (BOLC and ECCC) prepared you to provide terrain expertise to your unit and other units? Circle one answer.

	1	2	3	4	5
	No training given	Inadequate, or training was non-applicable	OK, but more needed	Good, Right amount	More than enough

1	2	3	4	5	Average
7	25	35	20	1	2.8

7. What skills do you believe need further emphasis in the engineer branch to enable you to succeed in providing terrain expertise to the field?

51	More training on TerraBase II and other automated terrain tools
44	More practice conducting terrain analysis
39	More practice conducting terrain-related reconnaissance
37	More exposure and interaction with 215D warrant officers and terrain teams
35	Greater understanding of digital terrain data and maps
35	Practice of terrain analysis for urban operations
26	Greater definition of the terrain expert role and responsibilities
26	Additional training tools available in the unit on terrain expert skills
25	More training on how to order and store digital and hardcopy terrain products

8. What other education or training have you received in the area of terrain analysis or understanding that has been helpful to you as an engineer officer?

27	Precommissioning training in ROTC, USMA, or Officer Candidate School
26	Unit-training on terrain analysis or TerraBase II
23	Experience at a CTC rotation(s)
23	None of the above
13	Undergraduate courses or degree in a terrain-related subject(s)
10	Civilian pursuits or jobs, such as hunting, surveying, or orienteering
7	Terrain-related training at another branch school
5	Training at NGS
1	Long-distance CD-ROM or web-based training on terrain visualization

9. Do you believe the training you have received in your military career has prepared you to be the terrain expert for the commander, staff, and BOS elements in your next assignment(s)?

Yes	No	Not Sure
29	37	22

10. Other comments?

[The following are the raw comments grouped by common subjects.]

Products
- Trafficability and line of sight (LOS) products are a big help.
- Satellite imagery is a success when it can be tied with a MCOO.
- Identify potential enemy locations with LOS analysis and imagery.

Terrain Teams
- The COO produced by the terrain team can be a great item.
- A positive technique at NTC was when the ABE used the detachment to help plan operations.
- It is often misused because the staff, to include the engineer officer, does not understand the capabilities or what to ask.

Hardware and Software Issues
- Need to learn to use the LOS tool on FBCB2.
- Most units do no that the assets to pull up and print digital products.
- Need a small color printer with a laptop to make handouts of key terrain, trafficability and range fans.
- Need more discussion on the limitations of digital systems/problems encountered.
- How much time does it take to produce DTSS and other terrain products?
- Need more software training for Falcon View, Mr. Sid, and ArcView.

Engineer Officer Responsibilities
- The ABE needs to push information to the TFEs.
- Need more practical exercises and hands on experience in BOLC/ECCC.
- Must know how to interpret products, not just make them.
- Get to know TerraBase II.
- It is important that the engineer officer listen to the senior terrain expert (for example a terrain analysis warrant officer) available to them.
- Know the capabilities of friendly and enemy weapon systems and how terrain impacts their effectiveness.
- BOLC is the beginning of training; battalion and company commanders need to ensure that the TFEs understand their responsibility to the IPB process.

Terrain Advice
- My unit did a poor job of assisting the BCT commander with terrain analysis.
- Engineers are not giving commanders terrain analysis and terrain effects on maneuverability.
- Best results accomplished when the terrain was related to actual battlefield effects.
- Inform staff of products the engineer can provide.
- Point out multiple vantage points of the objective during the OPORD brief.
- Most BOS commanders are excited about anything we can provide.
- Must know what products the commanders are looking for.

Miscellaneous
- Do reconnaissance--"boots-on-the-ground." Then you'll understand the real terrain out there.
- Units want trained second lieutenants with terrain expertise, but most of my training came from the unit.

APPENDIX C

CRITICAL TASK LISTS FOR THE 215D AND 81T SOLDIERS

<u>Critical Task List for 215D Terrain Analysis Warrant Officers</u>

ID NO	TASK NO	TASK TITLE	SKILL LEVEL
1	052-020-1001	Recommend changes to TOE, MTOE and TDA revisions	W1/W2
2	052-020-1002	Prepare unit status report	W1/W2
3	052-020-1003	Manage unit annual funding program	W1/W2
4	052-020-1004	Maintain property accountability	W1/W2
5	052-020-2001	Plan topographic support at the tactical level	W1/W2
6	052-020-2002	Plan topographic support at the joint/combined Level	W1/W2
7	052-020-2003	Plan crisis support operations	W1/W2
8	052-020-3001	Manage terrain analysis operations	W1/W2
9	052-020-3002	Manage reproduction of topographical products utilizing non-photolithography	W1/W2
10	052-020-3003	Exploit national sources	W1/W2
11	052-020-3004	Coordinate reproduction operations with supporting/supported units	W1/W2
12	052-020-3005	Manage remotely sensed imagery interpretation operations	W1/W2
13	052-020-3006	Request GI&S support	W1/W2
14	052-020-3007	Implement GI&S quality assurance procedures	W1/W2
15	052-020-3008	Implement MILSPECS into GI&S production	W1/W2
16	052-020-3009	Manage GI&S production	W1/W2
17	052-020-3010	Conduct rapid assessments of terrain line of sight and mobility	W1/W2
18	052-020-3011	Produce scaleable integrated digital terrain data (DTD) and TDAs	W1/W2
19	052-020-3012	Produce common portrayals of the physical characteristics of the battlefield	W1/W2
20	052-020-3013	Predict mobility	W1/W2
21	052-020-3014	Utilize mission planning and analysis system	W1/W2
22	052-020-4001	Manage GI&S database	W1/W2
23	052-020-4002	Distribute DTD in real-time or near real-time	W1/W2
24	052-020-4003	Collect DTD in real-time or near real-time	W1/W2
25	052-020-4004	Catalog DTD in real-time or near real-time	W1/W2
26	052-020-4005	Warehouse DTD in real-time or near real-time	W1/W2
27	052-020-4006	Update DTD in real-time or near real-time	W1/W2
28	052-020-4007	Synchronize data updates from various sources	W1/W2
29	052-020-4008	Track meta data for DTD	W1/W2
30	052-020-4009	Share DTD horizontally and vertically on the battlefield	W1/W2
31	052-020-4010	Transform DTD in real-time or near real-time	W1/W2
32	052-020-4011	Evaluate intended use of GI&S data	W1/W2
33	052-020-4012	Evaluate data currency and accuracy	W1/W2
34	052-020-4013	Implement data format compatibility with GI&S architecture	W1/W2
35	052-020-4014	Implement file transfer protocol	W1/W2
36	052-020-4015	Manage GI&S file servers	W1/W2

Critical Task List for 215D Terrain Analysis Warrant Officers (continued)

ID NO	TASK NO	TASK TITLE	SKILL LEVEL
37	052-020-4016	Maintain GI&S homepage	W1/W2
38	052-020-4017	Order standard NIMA GI&S data	W1/W2
39	052-020-4018	Order nonstandard GI&S data	W1/W2
40	052-020-4019	Maintain GI&S deployable databases	W1/W2
41	052-020-4020	Submit requests for information	W1/W2
42	052-020-4021	Rapidly connect to military civil and government geographic information data	W1/W2
43	052-020-5001	Determine the impact of GI&S operations on BOSs	W1/W2
44	052-020-5002	Integrate GI&S operations into BOSs	W1/W2
45	052-020-5003	Identify Army organizational structure	W1/W2
46	052-020-5004	Identify requirements of BOS	W1/W2
47	052-020-5005	Identify topographic requirements of intelligence preparation of the BOS	W1/W2
48	052-020-5006	Identify staff responsibilities	W1/W2
49	052-020-5007	Identify equipment/weapon system capabilities	W1/W2
50	052-020-5008	Produce tactical decision aids to support MDMP	W1/W2
51	052-020-5009	Identify requirements of command estimate process	W1/W2
52	052-020-5010	Identify requirements of intelligence estimate	W1/W2
53	052-020-5011	Identify topographic requirements of analysis of the AO	W1/W2
54	052-020-5012	Identify requirements of mission analysis	W1/W2
55	052-020-5013	Integrate environmental effects into tactical decision aids	W1/W2
56	052-020-5014	Conduct operational briefings	W1/W2
57	052-020-6001	Establish system architecture/connectivity	W1/W2
58	052-020-6002	Establish networking	W1/W2
59	052-020-6003	Maintain current topographic common hardware/software	W1/W2
60	052-020-6004	Maintain common operating equipment	W1/W2
61	052-020-6005	Identify requirements of account management and software licensing	W1/W2
62	052-020-7001	Plan topographic support at the strategic level	W3/W4
63	052-020-7002	Plan topographic support at the operational level	W3/W4
64	052-020-7003	Plan topographic support at the joint/combined level	W3/W4
65	052-020-2003	Plan crisis support operations	W3/W4
66	052-020-8001	Perform a digital color separate	W3/W4
67	052-020-8002	Create a cyan, magenta, yellow, and black (CMYK) negative	W3/W4
68	052-020-8003	Perform digital layout	W3/W4
69	052-020-8004	Exploit national sources	W3/W4
70	052-020-8005	Generate digital terrain data	W3/W4
71	052-020-8006	Generate digital elevation data	W3/W4
72	052-020-8007	Generate feature data	W3/W4
73	052-020-8008	Generate an ortho-rectified image data set	W3/W4
74	052-020-8009	Validate GI&S requirements	W3/W4

Critical Task List for 215D Terrain Analysis Warrant Officers (continued)

ID NO	TASK NO	TASK TITLE	SKILL LEVEL
75	052-020-8010	Prioritize GI&S requirements	W3/W4
76	052-020-8011	Prepare topographic orders and annexes	W3/W4
77	052-020-8012	Execute topographic missions	W3/W4
78	052-020-8013	Plan topographic missions	W3/W4
79	052-020-8014	Forecast personnel requirements	W3/W4
80	052-020-8015	Forecast material requirements	W3/W4
81	052-020-8016	Forecast equipment requirements	W3/W4
82	052-020-8017	Forecast training requirements	W3/W4
83	052-020-9001	Establish complex system architecture/connectivity	W3/W4
84	052-020-6002	Establish networking	W3/W4
85	052-020-6003	Maintain current topographic common hardware/software	W3/W4
86	052-020-6004	Maintain common operating environment	W3/W4
87	052-020-6005	Identify requirements of account management and software licensing	W3/W4

Critical Task List for 81T Non-Commissioned Officers and Soldiers

ID NO	TASK NO	TASK TITLE	SKILL LEVEL
1	052-245-1002	Extract EEGI--surface materials	1
2	052-245-1003	Extract EEGI--bridges	1
3	052-245-1004	Extract EEGI--railroads	1
4	052-245-1005	Extract EEGI--roads	1
5	052-245-1006	Extract EEGI--tunnels	1
6	052-245-1007	Extract EEGI--surface drainage	1
7	052-245-1008	Extract EEGI--airfields	1
8	052-245-1009	Extract EEGI--existing linear obstacles	1
9	052-245-1010	Extract EEGI--vegetation	1
10	052-245-1014	Extract EEGI--navigable inland waterways	1
11	052-245-1040	Extract features from hardcopy RSI	1
12	052-245-1071	Extract EEGI--surface configuration	1
13	052-245-1016	Produce a combined obstacles product digitally	1
14	052-245-1017	Produce a concealment overlay product digitally	1
15	052-245-1018	Produce a river crossing product digitally	1
16	052-245-1019	Produce a line of communications product digitally	1
17	052-245-1022	Produce a landing zone (LZ) product	1
18	052-245-1023	Produce a helicopter LZ (HLZ) product	1
19	052-245-1024	Produce a drop zone (DZ) product	1
20	052-245-1025	Produce a port analysis product	1
21	052-245-1029	Label interior features	1
22	052-245-1031	Produce a product template	1
23	052-245-1035	Scan a hard copy product	1
24	052-245-1042	Produce a cross country movement (CCM) product	1
25	052-245-1043	Rectify DTD on DTSS-B	1
26	052-245-1044	Mosaic DTD	1
27	052-245-1046	Produce image maps	1
28	052-245-1051	Digitize GI data	1
29	052-245-1052	Produce 3D fly through	1
30	052-245-1054	Perform image enhancements	1
31	052-245-1055	Produce single color overprints (SCO)	1
32	052-245-1058	Produce a visibility product	1
33	052-245-1061	Produce a mobility corridor product	1
34	052-245-1062	Prepare a MOUT product	1
35	052-245-1064	Perform digital change detection	1
36	052-245-1066	Perform image unsupervised classification	1
37	052-245-1072	Produce a combined obstacles product manually	1
38	052-245-1073	Produce a concealment overlay product manually	1
39	052-245-1074	Produce a river crossing product manually	1
40	052-245-1075	Produce a line of communications product manually	1
41	052-245-1081	Produce an off road speed product	1
42	052-245-1082	Produce an on road speed product	1

Critical Task List for 81T Non-Commissioned Officers and Soldiers (continued)

ID NO	TASK NO	TASK TITLE	SKILL LEVEL
43	052-245-1084	Perform basic map reading	1
44	052-245-1088	Perform terrain visualization briefing techniques	1
45	052-245-1089	Produce a cover product digitally	1
46	052-245-1090	Produce a cover product manually	1
47	052-245-1091	Produce a key terrain product digitally	1
48	052-245-1092	Produce a key terrain product manually	1
49	052-245-1033	Prepare DTSS-H for movement	1
50	052-245-1034	Prepare DTSS-H for operation	1
51	052-245-1036	Perform PMCS on DTSS equipment	1
52	052-245-1047	Prepare DTSS-L for operation	1
53	052-245-1048	Prepare DTSS-D for operation	1
54	052-245-1049	Prepare DTSS-L for movement	1
55	052-245-1050	Prepare DTSS-D for movement	1
56	052-245-1078	Operate The DTSS-H in a degraded mode	1
57	052-245-1079	Operate The DTSS-L in a degraded mode	1
58	052-245-1080	Operate The DTSS-D in a degraded mode	1
59	052-245-1083	Establish DTSS connectivity with ABCS	1
60	052-245-1094	Establish file directories and paths	1
61	171-145-0001	Send combat messages using FBCB2	1
62	171-145-0002	Prepare FBCB2 for operation	1
63	171-145-0006	Perform message management using FBCB2	1
64	171-145-0007	Send overlays using FBCB2	1
65	171-145-0008	Send reports using FBCB2	1
66	171-145-0011	Perform operator maintenance on FBCB2	1
67	171-145-0012	Perform shut-down procedures for FBCB2	1
68	052-245-1038	Import GI data	1
69	052-245-1039	Export GI data	1
70	052-245-1076	Install data on map server	1
71	052-245-1077	Integrate GI into the COP	1
72	052-245-1085	Create a database using the DTSS-B	1
73	052-245-1086	Update a database using the DTSS-B	1
74	052-245-1087	Disseminate a database using the DTSS-B	1
75	052-245-1093	Establish a GI database	1
76	052-245-2001	Maintain a GI database	2
77	052-245-2006	Produce an imagery index	2
78	052-245-2034	Order NIMA products	2
79	052-245-2035	Archive geospatial information and products	2
80	052-245-2036	Build a product data query	2
81	052-245-2037	Maintain digital data on the map server	2
82	052-245-2038	Update geospatial information	2
83	052-245-2040	Disseminate geospatial products	2
84	052-245-2039	Perform quality control checks and assurance on DTSS products	2

Critical Task List for 81T Non-Commissioned Officers and Soldiers (continued)

ID NO	TASK NO	TASK TITLE	SKILL LEVEL
85	052-245-2008	Supervise preparation of DTSS-H for operation	2
86	052-245-2018	Supervise PMCS on DTSS equipment	2
87	052-245-2019	Supervise preparation of DTSS-L for operation	2
88	052-245-2020	Supervise preparation of DTSS-D for operation	2
89	052-245-2021	Supervise preparation of DTSS-H for movement	2
90	052-245-2022	Supervise preparation of DTSS-L for movement	2
91	052-245-2023	Supervise preparation of DTSS-D for movement	2
92	052-245-2041	Load operating system	2
93	052-245-2042	Load network printer software	2
94	052-245-3003	Validate source materials	3
95	052-245-3005	Determine weather effects on operations	3
96	052-245-3006	Prepare analysis of the area of operations (AAO) product	3
97	052-245-3007	Conduct a terrain visualization brief	3
98	052-245-3008	Prepare a flood prediction product	3
99	052-245-3009	Prepare an avenue of approach product	3
100	052-245-3010	Plan GI production	3
101	052-245-3011	Update attribute categories for GI&S	3
102	052-245-3015	Determine supply requirements on GI&S project	3
103	052-245-3036	Produce amphibious beach LZ analysis product	3
104	052-245-3037	Perform image supervised classification	3
105	052-245-3043	Prepare a production schedule	3
106	052-245-3041	Supervise maintenance of the DTSS-B	3
107	052-245-3038	Supervise creating a database using the DTSS-B	3
108	052-245-3039	Supervise updating a database using the DTSS-B	3
109	052-245-3040	Supervise disseminating a database using the DTSS-B	3
110	052-245-3042	Integrate GI&S products into the MDMP	3
111	052-245-4002	Perform system administration on DTSS computer equipment	4
112	052-245-4003	Connect DTSS to LAN	4
113	052-245-4008	Establish user accounts and permissions on DTSS	4
114	052-245-4023	Maintain user accounts and permissions on DTSS	4
115	052-245-4024	Build a DTSS homepage	4
116	052-245-4025	Maintain DTSS products on the homepage	4
117	052-245-4016	Supervise quality assurance checks on DTSS products	4
118	052-245-4026	Determine GI&S requirements to support operations	4
119	052-245-4027	Integrate terrain visualization into the Army and joint planning process	4
120	052-245-4028	Coordinate actions with staff organizations	4
121	052-245-4029	Conduct equipment management	4
122	052-245-4030	Conduct personnel management	4
123	052-245-4031	Conduct budget management	4
124	052-245-4032	Conduct supply management	4
125	052-245-4033	Conduct training management	4

APPENDIX D

ESSENTIAL TASK LIST FOR THE 21B ENGINEER OFFICER

Essential SKBs Supporting "Assess"

ESSENTIAL TASKS	SKB	RANK	GEOSPATIAL FUNCTIONS	INSTITUTIONAL TRAINING	SOURCE(S)
Understand and coordinate the seven geospatial engineering functions at tactical echelons.	B	2LT +	Terrain Advice	NGS	FM 3-34.231; FM 7-15; ARTEP 5-335-65-MTP; STP 5-21II-MQS
Understand the roles of the engineer officer, terrain analysis warrant officer and topographic analyst.	K	2LT+	Terrain Advice	BOLC, ECCC, PCC, TV-CD	TPIO-TD
Understand the capabilities of an engineer terrain team and the DTSS.	K	2LT +	Data Manipulation and Exploitation	BOLC, ECCC, PCC, TV-CD, NGS	TPIO-TD; ARTEP 5-335-65-MTP; *Engineer* (1998 &1992); FM 3-34.221
Translate tactical requirements into GI&S products and analysis.	B	1LT+	Terrain Advice	BOLC, ECCC, PCC, TV-CD, NGS	*Engineer* (1996)
Understand the importance of military topography and map production to senior leaders.	K	CPT+	Terrain Advice	PCC, NGS	Prol; von Wahlde; Dunn
Understand the organization and capabilities of the DS corps topographic engineer company.	K	CPT +	Data Generation	BOLC, ECCC, TV-CD, NGS	FM 5-100
Integrate a terrain team in brigade operations.	K	CPT +	Terrain Advice	None	FM 3-34.221; *Engineer* (1998)
Understand the geospatial capabilities available from different services in Joint Operations	K	MAJ +	Data Collection	NGS	STP 21-III-MQS; FM 5-100
Understand how space-based systems can enhance warfighting capabilities at the tactical and higher levels of war in reconnaissance, position and navigation, and weather	K	MAJ +	Data Collection	NGS	STP 21-III-MQS
Understand what topographic support is available for MOOTW.	K	CPT +	Terrain Advice	TV-CD, NGS	FM 3-34.221
Understand map datums and scales.	K	2LT +	Data Collection	USMA, ECCC, PCC, TV-CD, NGS	*Engineer* (1999)
Understand the levels of accuracy provided by geodetic survey and GPS.	K	CPT +	Geodetic Survey	NGS	FM 3-34.231

Essential SKBs Supporting "Acquire"

ESSENTIAL TASKS	SKB	RANK	GEOSPATIAL FUNCTIONS	INSTITUTIONAL TRAINING	SOURCE(S)
Understand the different types of digital GI&S data and their military uses.	K	2LT +	Data Manipulation & Exploitation	BOLC, ECCC, PCC, TV-CD, NGS	TPIO-TD; *Engineer* (1999)
Understand the FD construct from NIMA.	K	2LT +	Data Collection	BOLC, PCC, NGS	*Engineer* (2002 & 2001)
Evaluate the availability of standard and nonstandard map products.	B	CPT +	Data Collection	BOLC, ECCC, TV-CD, NGS	FM 5-100
Request standard NIMA products through logistics channels or directly from DLA.	S	2LT +	Data Collection	BOLC, ECCC, TV-CD, NGS	FM 3-34.221; *Engineer* (1999)
Integrate nonstandard and non-US GI&S products into tactical databases.	K	CPT +	Database Management	ECCC	*Engineer* (1995)
Disseminate terrain analysis and other geospatial products.	B	CPT +	Database Management	NGS	*CTC Newsletter 99-12*; FM 3-34.221
Transmit essential terrain information to terrain teams for product update.	B	CPT +	Data Generation	None	FM 5-100
Understand the digital size of DTSS GI&S products made for ABCS use.	K	CPT +	Data Manipulation & Exploitation	None	TPIO-TD
Resolve differences between reports and products to render a single COP.	B	CPT +	Database Management	None	*Engineer* (2001 & 1999)
Track templated and known obstacles (friendly/enemy).	S	1LT +	Data Collection	None	FM 3-34.221
Establish IR for essential elements of terrain or engineer information.	K	CPT +	Data Collection	None	STP 5-21II-MQS ; ARTEP 5-335-65-MTP; FM 3-34.231; FM 3-34.221
Coordinate with the S2/G2 and S3/G3 for collecting terrain information.	B	CPT +	Data Collection	None	FM 3-34.231
Coordinate with the S2/G2 to define, prioritize, and request GI&S products.	B	MAJ +	Data Generation	None	FM 3-34.231
Prepare a risk assessment when the lack of geospatial information creates uncertainty.	B	CPT+	Terrain Advice	None	*Engineer* (1996)
Direct engineer reconnaissance missions.	B	2LT +	Data Collection	BOLC	STP 5-21II-MQS ; FM 5-100; FM 5-170
Conduct a reconnaissance.	S	2LT +	Data Collection	BOLC	STP 21-I-MQS ; FM 5-100; FM 5-170; ARTEP 5-335-65-MTP
Submit both verbal and written patrol reports as required by STANAG 2003.	B	2LT +	Data Collection	BOLC	ARTEP 5-335-65-MTP
Record terrain information daily.	B	2LT +	Data Collection	None	Swift; Dunn

Essential SKBs Supporting "Analyze"

ESSENTIAL TASKS	SKB	RANK	GEOSPATIAL FUNCTIONS	INSTITUTIONAL TRAINING	SOURCE(S)
Conduct terrain analysis.	S	2LT +	Terrain Analysis	ROTC, USMA, BOLC, ECCC, TV-CD	STP 21-I-MQS; Sun Tzu; FM 3-0; FM 3-34.221; FM 5-100; FM 34-130; FM 5-33; *CTC Trends 96-12*
Perform map reading.	S	2LT +	Navigation	ROTC, USMA, BOLC, ECCC	STP 21-1-SMCT
Measure distance on a map.	S	2LT +	Navigation	ROTC, USMA, BOLC, ECCC	STP 21-1-SMCT
Identify topographic symbols on a map.	S	2LT +	Navigation	ROTC, USMA, BOLC, ECCC	STP 21-1-SMCT; Maclean; Musham
Identify terrain features on a map.	S	2LT +	Navigation	ROTC, USMA, BOLC, ECCC	STP 21-1-SMCT
Evaluate terrain 360 degrees from desired locations.	B	2LT +	Terrain Analysis	None	Swinton
Understand enemy weapon's capabilities.	K	2LT +	Terrain Analysis	BOLC, ECCC	Swinton
Use terrain to deceive the enemy.	S	1LT +	Terrain Analysis	None	Swinton
Identify weather impacts on the terrain.	K	2LT +	Terrain Analysis	BOLC, ECCC	ARTEP 5-335-65-MTP
Understand how the terrain affects unit camouflage operations.	K	1LT +	Terrain Analysis	None	ARTEP 5-335-65-MTP
Understand military geography for different regions.	K	2LT +	Terrain Analysis	None	*Engineer* (1992)
Understand geospatial information requirements of each BOS.	K	MAJ +	Terrain Advice	ECCC, TV-CD	STP 21-III-MQS; FM 34-130; Laporte
Conduct terrain analysis for urban environments.	S	1LT +	Terrain Analysis	None	FM 3-0; FM 3-34.221; Johnson; Burleson; Kramer
Conduct terrain analysis for peacekeeping operations.	S	CPT+	Terrain Analysis	None	*Engineer* (1995)
Prepare a MCOO.	S	2LT +	Terrain Analysis	BOLC, ECCC, TV-CD	FM 5-100
Use PC-based terrain analysis tools, such as TerraBase II, to create TDAs.	S	2LT +	Data Manipulation & Exploitation	BOLC, ECCC, TV-CD, NGS	*CTC Bulletin 02/17*; *Engineer* (1998, 1997, & 1995); Laporte
Prepare engineer estimates to include geospatial engineering capabilities.	S	CPT +	Terrain Analysis	BOLC, ECCC	STP 5-21II-MQS; FM 3-34.221

Essential SKBs Supporting "Analyze" (continued)

ESSENTIAL TASKS	SKB	RANK	GEOSPATIAL FUNCTIONS	INSTITUTIONAL TRAINING	SOURCE(S)
Provide input to IPB.	B	CPT +	Terrain Analysis	BOLC, ECCC	STP 5-21II-MQS; FM 34-130; FM 5-100; FM 3-34.221; STP 21-III-MQS
Perform grid coordinate conversions.	S	2LT +	Data Manipulation & Exploitation	BOLC, ECCC, NGS	*Engineer* (1995)
Visualize the terrain.	S	1LT +	Terrain Analysis	BOLC, ECCC, PCC	*Engineer* (1996)
Understand geospatial engineering's role in assured mobility.	K	CPT +	Terrain Advice	None	FM 3-34.221
Provide bridge classification maps.	S	CPT +	Cartographic Production & Reproduction	None	FM 5-100
Execute target-folder battle drills.	S	CPT +	Data Generation	BOLC, ECCC, TV-CD	ARTEP 5-335-65-MTP

Essential SKBs Supporting "Advise"

ESSENTIAL TASKS	SKB	RANK	GEOSPATIAL FUNCTIONS	INSTITUTIONAL TRAINING	SOURCE(S)
Understand the BTRA capabilities embedded in ABCS platforms.	K	CPT +	Data Manipulation & Exploitation	None	*Engineer* (2002)
Understand the capabilities of JMTK as a component of ABCS.	K	CPT +	Data Manipulation & Exploitation	TV-CD	*Engineer* (2002)
Translate GI&S into tactical terms for the warfighter.	B	1LT +	Terrain Advice	BOLC, ECCC, PCC, TV-CD	*Engineer* (1996)
Brief terrain effects.	B	2LT +	Terrain Advice	ROTC, USMA, BOLC, ECCC, TV-CD	FM 5-100
Advise the commander on the use of terrain for combat operations.	B	2LT +	Terrain Advice	BOLC, ECCC	STP 5-21II-MQS; FM 5-100; ARTEP 5-335-65-MTP; *CTC Newsletter 96-12*; *CTC Trends* 97-16; *(CTC Trends* 01-2; *Engineer* (2000 & 1989)
Assist the maneuver commander with terrain visualization.	B	CPT +	Terrain Advice	BOLC, ECCC, TV-CD	TRADOC Pam 525-41
Provide geospatial engineering advice to S4/G4 for MSR & logistics operations.	B	CPT +	Terrain Advice	None	FM 3-34.221
Provide the status of infrastructure for contingency operations.	S	CPT +	Data Generation	None	FM 5-100; FM 3-34.221

Essential SKBs Supporting "Apply"

ESSENTIAL TASKS	SKB	RANK	GEOSPATIAL FUNCTIONS	INSTITUTIONAL TRAINING	SOURCE(S)
Determine the grid coordinates of a point on a military map.	S	2LT +	Navigation	ROTC, USMA, BOLC, ECCC	STP 21-1-SMCT
Orient a map to the ground by map terrain association.	S	2LT +	Navigation	ROTC, USMA, BOLC	STP 21-1-SMCT; Swift
Determine a location on the ground by terrain association.	S	2LT +	Navigation	ROTC, USMA, BOLC	STP 21-1-SMCT
Determine a magnetic azimuth using a lensatic compass.	S	2LT +	Navigation	ROTC, USMA, BOLC	STP 21-1-SMCT
Navigate using a map and a compass.	S	2LT +	Navigation	ROTC, USMA, BOLC	STP 21-1-SMCT
Determine direction without a compass or GPS.	S	2LT +	Navigation	ROTC, USMA, BOLC	STP 21-1-SMCT
Navigate using GPS equipment.	S	2LT +	Navigation	USMA, NGS	
Identify the critical terrain elements for a unit defensive position	B	2LT +	Terrain Analysis	BOLC	ARTEP 5-335-65-MTP
Supervise site selection and layout	B	2LT +	Terrain Analysis	None	STP 5-21II-MQS; Swinton
Use a digital SA overlay to conduct a map reconnaissance.	S	CPT +	Data Manipulation & Exploitation	None	ARTEP 5-335-65-MTP
Use a digital SA overlay to conduct a map orientation.	S	CPT +	Data Manipulation & Exploitation	None	ARTEP 5-335-65-MTP
Preparing the topographic annex or appendix to tactical plans and orders.	B	MAJ +	Terrain Advice	NGS	FM 3-34.231; FM 3-34.221

GLOSSARY

Army Battle Command System (ABCS). An integrated network of battlefield automated information systems, providing a seamless C2 capability from the strategic echelon to the foxhole. Its purpose is to help commanders obtain optimal, near-real-time access to the commander's critical information requirements (CCIR) through force-level databases (FM 3-34.230 2000, 4-2).

Assured Mobility. Actions that guarantee the force commander the ability to maneuver where and when he desires without interruption or delay to achieve his intent (FM 3-34.221 2002, 1-2).

Battlefield Operating System (BOS). A listing of critical tactical activities: intelligence, maneuver, fire support, mobility and survivability, air defense, combat service support, and command and control (FM 101-5-1 1997, 1-18).

Battlefield Terrain and Reasoning Awareness (BTRA). This allows friendly forces the ability to template potential obstacles and locations of where the enemy might place obstacles (Fowler and Johnston 2002, 13).

Battlespace. The conceptual physical volume of space in which the commander seeks to dominate the enemy. It encompasses three dimensions and is influenced by the operational dimensions of time, tempo, depth, and synchronization (FM 101-5-1 1997, 1-18).

Cartographic Production and Reproduction. Cartographic production is the preparation of drawings and materials for special purpose graphics and revision of existing cartographic and imagery products. Reproduction is the process of creating hard-copy maps and geospatial information (GI) products from original drawings, reproduction materials (repro mats), or images (FM 3-34.221 2002, 3-7).

Commercial Joint Mapping Tool Kit (C/JMTK). It is a standardized, commercial, comprehensive tool kit of software components for the management, analysis, and visualization of map and map-related information produced by Northrup Grumman Information Technology. C/JMTK will replace the government-owned software package, JMTK (TASC 2003).

Data Collection. The GI database development begins during an operation's pre-deployment phase in order to gain maximum knowledge of the potential AO and AOI. Following deployment, enriched data will be collected using all available information (FM 3-34.221 2002, 3-13).

Data Generation. The creation of GI&S data and products to fill in voids in existing databases and to intensify the detail of the AO and AOI with mission specific data sets.

Database Management. The acquisition, manipulation, formatting, storage, and distribution of all hard-copy and digital data and products. GI databases include old data, new data, accurate data, qualified data, and multi-formatted hard-copy and digital data (FM 3-34.221 2002, 3-7).

Data Manipulation and Exploitation. The shaping of existing GI data into tactical decision aids and analysis to facilitate terrain analysis in the planning processes and to fulfill BOS GI requirements.

Digital Topographic Support System (DTSS). The topographic-engineer and topographic-analysis component of ABCS that provides critical, timely, accurate, and analyzed digital and hard-copy mapping products to the battle commander for terrain visualization. It is used by engineer topographic teams and companies (FM 3-34.230 2000, 4-5).

Force XXI Battle Command Brigade and Below (FBCB2). A battle command information support system supported by existing and emerging communications, sensors, and electrical power sources. It provides command, control, and situational awareness capabilities to all echelons of the task force using a mix of ruggedized computers, radio transmitters, position-navigation equipment, and tactical local area network hardware (FM 3-34.221 2002, C-1).

Geodetic Survey. The process of determining the relative positions of points on, above, or beneath the earth's surface by using traditional or satellite-based measurement systems. Survey teams are found at Corps and higher and support terrain platoons/teams, field artillery, Army aviation, air defense artillery, intelligence, chemical, armor, combat service support (CSS), signal, US Air Force (USAF), and NIMA (FM 3-34.221 2002, 3-7).

Geospatial Engineering. The collection, development, dissemination and analysis of positionally accurate terrain information that is tied to some earth reference, to provide mission tailored data, tactical decision aids and visualization products that define the character of the zone for the maneuver commander (FM 3-34 2003, 4-9).

Imagery. Data collected in a particular limited region or subset of the electromagnetic spectrum that represents an object, scene, or map. Aerial photography and satellite images are types of imagery (National Imagery and Mapping Agency 1995, F-2).

Joint Mapping Tool Kit (JMTK). A software package that consists of five components: Analysis (JMA), Visualization (JMV), Spatial Database Management (JMS), Imagery (JMI), and Utilities (JMU) (National Imagery and Mapping Agency 2003).

Maneuver Control System-Engineer (MCS-E). The automated decision support and management element that is tied into the DTSS and rest of ABCS to support the maneuver commander and provide the engineer commander with rapid answers to otherwise time-consuming, manual calculations (Aadland and Allen 2002, 9).

Objective Force. The Army's future full-spectrum force: organized, manned, equipped and trained to be more strategically responsive, deployable, agile, versatile, lethal, survivable and sustainable across the entire spectrum of military operations from Major Theater Wars through counter terrorism to Homeland Security. It is designed to be more strategically responsive and dominant at every point on the spectrum of military operations than the Legacy Force (White Paper 2002, iv).

Stryker Brigade Combat Team (SBCT). A brigade designed to optimize effectiveness and balance the traditional domains of lethality, mobility, and survivability with responsiveness, deployability, sustainability, and a reduced in-theater footprint. It is a full-spectrum combat force with the core qualities of mobility and decisive action through dismounted infantry assault (FM 3-34.221 2002, 1-1).

Terrain Analysis. The collection, analysis, evaluation, and interpretation of geographic information on the natural and man-made features of the terrain, combined with other relevant factors, to predict the effect of the terrain on military operations. (FM 101-5-1 1997, 1-153). Terrain analysis is conducted at two distinct levels of detail: general terrain analysis conducted by staffs using the basic principles of OCOKA, and detailed terrain analysis conducted by skill technicians (215D warrant officers and 81T soldiers) using automated systems and specialized training.

Terrain Visualization. The process through which a commander sees the terrain and understands its impact on the operation in which he is involved. It is the subjective evaluation of the terrain's physical attributes as well as the physical capabilities of vehicles, equipment, and personnel that must cross over and occupy the terrain. Terrain visualization is a component of battlefield visualization. Engineers are responsible for providing the means to achieve terrain visualization (FM 3-34.221 2002, 1-5).

Topography. The technique of describing, graphically representing, and surveying the exact physical features of a place or region on a map.

Unit of Action (UA). The tactical warfighting echelons of the Objective Force (White Paper 2002, 18).

Unit of Employment (UE). The basis of combined arms air-ground task forces. They resource and execute combat operations; designate objectives; coordinate with multi-service, interagency, multinational and non-governmental activities; and employ long range fires, aviation and sustainment. They also provide C4ISR and tactical direction to UAs (White Paper 2002, 18).

REFERENCE LIST

Aadland, Anders B., and James Allen. 2002. Engineer white paper--into the objective force. *Engineer,* April, 4-10.

Arnold, Edward. 1997. Being a terrain visualization expert. *Engineer,* March, 22-23.

ARTEP 5-335-65-MTP. 2000. See US Department of the Army. 2000a.

Bell, Stephen. 1999. Engineer terrain analysis. *CTC Newsletter* 99-12 (October): 39-42

Burleson, Willard M., III. 2000. Mission analysis during future military operations on urbanized terrain. M.A. thesis, Command and General Staff College.

Chamberlain, E. W., Kevin Williams, and Mario Perez. 1998. Engineers in the 21st century. *Engineer*, April, 19-21.

Chaney, Richard. 1998. Colonel, you're fired! *Engineer,* February, 15-17.

Clausewitz, Carl von. 1976. *On War*. Edited and Translated by Michael Howard and Peter Paret. Princeton, New Jersey: Princeton University Press.

Collins, John. 1998. *Military Geography for Professionals and the Public.* Washington, DC: National Defense University Press.

Crawford, Kenneth. 1998. Engineer support to engagement area development. *Engineer*, July, 40-45.

CTC Newsletter 96-12. 1996. See US Army Center for Army Lessons Learned. 1996b.

CTC Newsletter 98-10. 1998. See US Army Center for Army Lessons Learned. 1998.

CTC Newsletter 99-12. 1999. See US Army Center for Army Lessons Learned. 1999c.

CTC Newsletter 99-16. 1999. See US Army Center for Army Lessons Learned. 1999d.

CTC Trends 96-9. 1996. See US Army Center for Army Lessons Learned. 1996a.

CTC Trends 97-16. 1997. See US Army Center for Army Lessons Learned. 1997a.

CTC Trends 97-19. 1997. See US Army Center for Army Lessons Learned. 1997b.

CTC Trends 99-7. 1999. See US Army Center for Army Lessons Learned. 1999a.

CTC Trends 99-10. 1999. See US Army Center for Army Lessons Learned. 1999b.

CTC Trends 01-02. 2001. See US Army Center for Army Lessons Learned. 2001.

CTC Trends 02-17. 2002. See US Army Center for Army Lessons Learned. 2002b.

DA PAM 350-58. 1994. See US Department of the Army. 1994c.

Dunn, Dr. James. 2001. Army engineers in the west. *Engineer,* August, 52-57.

Dyke, Carl van. 1990. *Russian Imperial Military Doctrine and Education, 1832-1914.* Westport, CT: Greenwood Press.

Erwin, Ralph. 2001. Disseminating digital terrain data to warfighters. *Engineer,* May, 16-17.

Fahey, John. From the president. *National Geographic,* January 2003, xxi.

Fitzpatrick, John C. 1931. *Writings of George Washington from the Original Manuscript Sources, 1745-1799.* Washington, DC: Government Printing Office.

Flowers, Robert B. 1999. Prioritization paper for engineer future capabilities. Fort Leonard Wood, MO: US Army Maneuver Support Center.

FM 34-130. 1994. See US Department of the Army. 1994b.

FM 5-10. 1995. See US Department of the Army. 1995.

FM 5-100. 1996. See US Department of the Army. 1996b.

FM 17-95. 1996. See US Department of the Army. 1996c.

FM 101-5. 1997. See US Department of the Army. 1997b.

FM 101-5-1. 1997. See US Department of the Army. 1997c.

FM 90-13. 2000. See US Department of the Army. 2000b.

FM 3-34.230. 2000. See US Department of the Army. 2000c.

FM 3-0. 2001. See US Department of the Army. 2001c.

FM 3-34.221. 2002. See US Department of the Army. 2002a.

FM 7-0. 2002. See US Department of the Army. 2002b.

FM 7-15. 2002. See US Department of the Army. 2002c.

FM 3-34. 2003. See US Department of the Army. 2003.

Fowler, Mike, and Gary Johnston. 2002. Assured mobility--the path to the future. *Engineer,* April, 11-15.

Gill, Clair. 1996. Terrain visualization: the challenge for the whole team. *Engineer,* April, 11-15.

Granger, Sonja, Geospatial Engineering Instructor. 2003. Interview by author, 13 February, Fort Leonard Wood, Missouri. Phone interview. Command and General Staff College, Fort Leavenworth.

Griffith, Samuel B. 1963. *Sun Tzu, The Art of War.* Oxford, United Kingdom: Oxford University Press.

Hooper, Earl, Chris Morken, and Brian Murphy. 2001. Geospatial engineering: a rapidly expanding engineer mission. *Engineer,* May, 14-15.

Hooper, Earl, and Mark Adams. 1998. TerraBase II, version 3.0 - supporting the terrain visualization expert. *Engineer,* November, 30-32.

Hotchkiss, Jedediah. 1973. *Make Me a Map of the Valley.* Dallas, Texas: Southern Methodist University Press.

Kennedy, Daniel. 1998. *Surveying the Century.* Westphalia, Missouri: Westphalia Publishing.

Kirby, Russ. 1997. Introducing terrabase II. *Engineer,* August, 38-39.

Kirby, Russ, Geospatial Engineer Officer, TPIO-TD. 2003. Interview by author, 15 January, Fort Leonard Wood, Missouri. Phone interview. Command and General Staff College, Fort Leavenworth.

Kramer, Chris. 2002. Terrain analysis considerations. Fort Leonard Wood, MO: US Army Engineer School.

Laporte, Leon J., and David F. Melcher. 1997. Terrain visualization. *Military Review* 5 (September): 75-80.

Light, Ron. 1999. CTC notes. *Engineer,* November, 57-60.

Maclean, Norman F., and Everett C. Olsen. 1943. *Manual for Instruction in Military Maps and Aerial Photography.* New York: Harper and Brothers Publishers.

McGinley, Shawn. 1999. Assistant brigade engineer (ABE) proficiency. *CALL Newsletter* 99-12 (October): 78-79.

Medby, Jamison, and Russell Glenn. 2002. *StreetSmart: Intelligence Preparation of the Battlefield for Urban Operations.* Santa Monica, CA: RAND Corporation.

Metzer, Terri. 1992. Terrain analysis for desert storm. *Engineer,* February, 22.

Musham, Harry A. 1944. *The Technique of the Terrain.* New York: Reinhold Publishing Corporation.

National Imagery and Mapping Agency. 1995. *Multispectral User's Guide.* Washington DC: Government Printing Office.

_____. 2003. JMTK. National Imagery Mapping Agency. Database on-line. Available from: http://www.jmtk.org/. Accessed 9 April 2003.

O'Sullivan, Patrick. 1991. *Terrain and Tactics.* New York, New York: Greenwood Press.

Pierce, William. 2001. Going, going, gone . . . bidding farewell to the 1:50,000 scale topographic line map. *Engineer,* May 2001, 10-13.

Piek, Joseph, and Dirk Plante. 1998. Terrain analysis as a combat multiplier just got better. *Engineer,* April 1998, 20-22.

Plante, Dirk. 1999. What every warrior should know about maps. *Engineer,* November, 38-40.

Prol, Elbeurtus. 2002. Robert erksine. Database on-line. Available from http://www.ringwoodnj.net/erskine1.htm. Accessed 29 April 2003.

Prude, Forrest. 1999. Combat heavy and combat support equipment platoons: supporting brigades forward. *CTC Newsletter* 99-12 (October): 33-36.

Putnam, John. 1943. *Map Interpretation with Military Application.* New York: McGraw-Hill.

Reminger, Bruce H. 1983. Restructuring SC 21. *Engineer* 4: 35.

Rensema, Tim, Craig Erickson, and Steve Herda. 2000. GIS - the bridge into the twenty-first century. *Engineer,* April, 34-39.

Rohr, Konrad. 1956. The soldier and the map. *Military Review* 35, no. 7: 87-91.

Schubert, Frank N. 1980. *Vanguard of Expansion: Army Engineers in the Trans-Mississippi West 1819-1879.* Washington DC: Historical Division, Office of Administrative Services, Government Printing Office.

Simutis, Zita M., and Helena F. Barsam. 1982. *Terrain Visualization by Soldiers.* Alexandria, Virginia: US Army Research Institute for the Behavioral and Social Sciences.

Snyman, Eugene, and Kenneth Bergman. 2002. Engineer digital command and control. *Engineer,* April, 24-26.

Starke, Jeffrey, Instructor at USMA. 2003. Interview by author, 6 March, Kansas. Electronic mail. Command and General Staff College, Fort Leavenworth.

Stewart, Jeb. 1999. Engineers, army after next, and military operations in urban terrain. *Engineer,* March, 17-19.

STP 21-III-MQS. 1993. See US Department of the Army. 1993b.

Swift, Eben. 1897. The lyceum at fort agawam. *Journal of the Military Service Institution of the United States* XX, no. LXXXVI.

Swinton, E. D. 1986. *The Defence of Duffer's Drift.* Wayne, New Jersey: Avery Publishing Group Inc.

TASC. 2003. Commercial/Joint Mapping Tool Kit. Northrup Grumman Information Technology. Database on-line. Available from: http://www.cjmtk.com/data.pdf. Accessed 9 April 2003.

Tatro, Kenneth, Terrain Technician TPIO-TD. 2002. Interview by author, 18 November, Fort Leonard Wood, Missouri. Notes. Command and General Staff College, Fort Leavenworth.

Traas, Adrian George. 1993. *From the Golden Gate to Mexico City: Topographic Engineers in the Mexican War, 1846-1848.* Washington, DC: Government Printing Office, Center of Military History.

TRADOC Pam 525-41. 1997. See US Department of the Army. 1997d.

Treleaven, David L. 1995. We're all terrain experts. *Engineer,* February, 8-11.

Tupper, Steve, Deputy Director TPIO-TD. 2003. Interview by author, 27 January, Fort Leavenworth, Kansas. Notes. Command and General Staff College, Fort Leavenworth.

Turabian, Kate L. 1996. *A Manual for Writers.* 6th ed. Chicago, Illinois: University of Chicago Press.

Tzu, Sun. 2002. *The Art of War*. Trans. Lionel Giles. Dover ed. Harrisburg, PA: The Military Service Publishing Company, 1944; Mineola, NY: Dover Publications, Inc.

US Army Center for Army Lessons Learned. 1996a. Bulletin Number 96-9. *CTC Trends, Joint Readiness Training Center*. Fort Leavenworth: US Army Combined Arms Center.

_____. 1996b. Bulletin Number 96-12. *CTC Newsletter, National Training Center*. Fort Leavenworth: US Army Combined Arms Center.

_____. 1997a. Bulletin Number 97-16. *CTC Trends, National Training Center*. Fort Leavenworth: US Army Combined Arms Center.

_____. 1997b. Bulletin Number 97-19. *CTC Trends, Joint Readiness Training Center*. Fort Leavenworth: US Army Combined Arms Center.

_____. 1998. Bulletin Number 98-10. *CTC Newsletter, Joint Readiness Training Center*. Fort Leavenworth: US Army Combined Arms Center.

_____. 1999a. Bulletin Number 99-7. *CTC Trends, Joint Readiness Training Center*. Fort Leavenworth: US Army Combined Arms Center.

_____. 1999b. Bulletin Number 99-10. *CTC Trends, Joint Readiness Training Center*. Fort Leavenworth: US Army Combined Arms Center.

_____. 1999c. Bulletin Number 99-12. *CTC Newsletter, National Training Center*. Fort Leavenworth: US Army Combined Arms Center.

_____. 1999d. Bulletin Number 99-16. *CTC Trends, Joint Readiness Training Center*. Fort Leavenworth: US Army Combined Arms Center.

_____. 2001. Bulletin Number 01-2. *CTC Trends, Joint Readiness Training Center*. Fort Leavenworth: US Army Combined Arms Center.

_____. 2002a. Bulletin Number 02-5. *CTC Trends*. Fort Leavenworth: US Army Combined Arms Center.

_____. 2002b. Bulletin Number 02-17. *CTC Trends*. Fort Leavenworth: US Army Combined Arms Center. Database on-line. Available from: http://call.army.mil/Products/CTC_BULL/02-17/toc.htm. Accessed 8 April 2003.

US Department of the Army. 1988. Field Manual (FM) 25-100, *Training in Units*. Washington DC: Government Printing Office.

_____. 1990a. FM 5-33, *Terrain Analysis*. Washington DC: Government Printing Office.

_____. 1990b. Soldier's Training Publication (STP) 21-I-MQS, *Military Qualification Standards I Manual of Common Tasks (Precommissioning Requirements)*. Washington DC: Government Printing Office.

_____. 1991. STP 5-21II-MQS, *Military Qualification Standards II Engineer (21) Company Grade Officer's Manual*. Washington DC: Government Printing Office.

_____. 1993a. FM 5-71-100, *Division Engineer Combat Operations*. Washington DC: Government Printing Office.

_____. 1993b. STP 21-III-MQS, *Military Qualification Standards III Leader Development for Majors and Lieutenant Colonels*. Washington DC: Government Printing Office.

_____. 1994a. FM 5-7-30, *Brigade Engineer and Engineer Company Combat Operations (Airborne, Air Assault, Light)*. Washington DC: Government Printing Office.

_____. 1994b. FM 34-130, *Intelligence Preparation of the Battlefield*. Washington DC: Government Printing Office.

_____. 1994c. Department of the Army Pamphlet (DA PAM) 350-58, *Leader Development for America's Army*. Washington DC: Government Printing Office.

_____. 1994d. STP 21-1-SMCT, *Soldier's Manual of Common Tasks*. Washington DC: Government Printing Office.

_____. 1995. FM 5-10, *Combat Engineer Platoon*. Washington DC: Government Printing Office.

_____. 1996a. FM 5-71-2, *Task Force Engineer Combat Operations*. Washington DC: Government Printing Office.

_____. 1996b. FM 5-100, *Engineer Operations*. Washington DC: Government Printing Office.

_____. 1996c. FM 17-95, *Cavalry Operations*. Washington DC: Government Printing Office.

_____. 1997a. FM 5-71-3, *Brigade Engineer Combat Operations (Armored)*. Washington DC: Government Printing Office.

_____. 1997b. FM 101-5, *Staff Organization and Operations*. Washington DC: Government Printing Office.

_____. 1997c. FM 101-5-1, *Operational Terms and Graphics*. Washington DC: Government Printing Office.

_____. 1997d. Training and Doctrine Command (TRADOC) Pamphlet 525-41, *Topographic Support for Terrain Visualization*. Fort Monroe, Virginia: Government Printing Office.

_____. 2000a. ARTEP 5-335-65-MTP, *Mission Training Plan - Engineer Company; Engineer Company, Engineer Battalion, Heavy Division/Corps (Mech); Engineer Company, Heavy Separate Brigade/ACR*, Washington DC: Government Printing Office.

_____. 2000b. FM 90-13, *River Crossing Operations*. Washington DC: Government Printing Office.

_____. 2000c. FM 3-34.230, *Topographic Operations*. Washington DC: Government Printing Office.

_____. 2001a. FM 1, *The Army*. Washington DC: Government Printing Office.

_____. 2001b. FM 3-25.26, *Map Reading and Land Navigation*. Washington DC: Government Printing Office.

_____. 2001c. FM 3-0, *Operations*. Washington DC: Government Printing Office.

_____. 2001d. FM 3-90, *Tactics*. Washington DC: Government Printing Office.

_____. 2002a. FM 3-34.221, *Engineer Operations--Stryker Brigade Combat Team*. Washington DC: Government Printing Office.

_____. 2002b. FM 7-0, *Training in Units*. Washington DC: Government Printing Office.

_____. 2002c. FM 7-15, *The Army Universal Task List*. Washington DC: Government Printing Office.

_____. 2002e. The army white paper: concept for the objective force. Department of the Army. Database on-line. Available from: http://www.army.mil/features/WhitePaper/ObjectiveForceWhitePaper.pdf. Accessed 8 April 2003.

_____. 2003. FM 3-34 (Draft), *Engineer Operations*. Washington DC: Government Printing Office.

US Department of Defense. 1999. Joint Publication (JP) 2-03, *Joint Tactics, Techniques, and Procedures for Geospatial Information and Services Support to Joint Operations.* Washington DC: Government Printing Office.

Wahlde, Peter von. 1960. Russian military reform: 1862-1874. *Military Review* 39, no. 10: 1-19.

White Paper. 2002. See US Department of the Army. 2000e.

Wiersma, William. 1991. *Research Methods in Education: An Introduction,* 5th ed. University of Toledo. Boston, MA: Allyn and Bacon.

Williams, Morris. 1982. *The American Heritage Dictionary,* 2nd ed. Boston, MA: Houghton Mifflin Company.

Winters, Harold. 1998. *Battling the Elements.* Baltimore, MD: John Hopkins University Press.

Wright, Edward. 1992. Topographical challenge of desert shield and desert storm. *Military Review* 72, no. 3: 1-12.

www.ingramcontent.com/pod-product-compliance
Lightning Source LLC
Chambersburg PA
CBHW081132170526
45165CB00008B/2647